THE ILLUSTRATED ENCYCLOPEDIA OF CACTI & OTHER SUCCULENTS

THE ILLUSTRATED ENCYCLOPEDIA OF CACTI & OTHER SUCCULENTS

by J. Říha and R. Šubík

edited by Gillian and Kenneth A. Beckett

SELECT EDITIONS

Text and photographs by J. Říha and R. Šubík

Translated by D. Hábová
Designed by F. Prokeš
Line drawings by J. Mašek

This edition published 1992 by
The Promotional Reprint Co Ltd,
Chelsea Chambers, 262 A Fulham Road,
London SW10 9EL

ISBN 1 8564 8066 6
Printed in Czechoslovakia by SPEKTRUM, Brno
3/11/08/51-02

CONTENTS

WHAT ARE SUCCULENTS?

Vast areas of the surface of our planet, millions of square kilometres, suffer from a shortage of water. They have a low rainfall, irregularly distributed throughout the year, and lack permanent rivers and streams. Such regions are called deserts. In such areas, plants have evolved which show many adaptations to these adverse conditions. They are generally named xerophytes (*xeros* — dry), because they live in a dry environment. They can also be called xerophilous, literally meaning drought loving. If you set out to African or American deserts to see them, you will be amazed by their toughness. The scorching sun burns the stony ground, the temperature is about 40° C and you cannot rest on the roasting stones. If you camp there for the night, cold becomes the enemy. The temperature drops sharply after sunset, you will be chilled in your sleeping bag, and in the morning, you may find your so precious water frozen in the jerrican. But if you look arround carefully, you will discover tiny living plants scattered among the stones, many almost buried in the ground and often looking so like the surrounding stones and sand that the casual visitor would probably walk right over them without realizing their existence.

Not all these xerophilous plants are succulents, only those which have thick and fleshy vegetative parts (leaves or stems). When crushed, we can see they contain a watery liquid or juice. They store this water in their tissues and it allows them to resist the extreme desert conditions and survive in places only rarely visited by man. Sometimes it is very difficult to draw a line between succulent plants (*succus* — juice in Latin), and non-succulent plants (mesophytes) because both have evolved from the same beginnings and so have the same basic organs and structure.

MAN AND SUCCULENTS

Man has known and used succulents since ancient times. The excavations in Mexican Tehuacán prove the use of cactus fruits and seeds by cultures dating from many centuries before Christ. There are documents from the conquest of Latin America in the 16th century which mention the use of these plants by the natives. No wonder then that *tuna* and *nopalitos,* fruits and young segments of prickly pears (opuntias) are often seen in Mexican markets. Aloe flowers are used elsewhere in the world, and the seeds of many cacti are edible. Both fruits and parts of fleshy, succulent stems are candied and dried, or fermented, and eaten as jams and marmalades *(queso de tuna).* Beverages made of agave are very popular (*agua miel,* honey water), or in fermented form such as the alcoholic *pulque,* and the hard spirits *tequila* and *mezcal.* Leaves, stems, bast, spines and solid fibres find a wide range of application in building and basket making. The leaves of yucca, nolina, and many others, are used for roofing, in basket weaving or making of hats and rugs, and the tough leaves of some agave species were until recently made into sandals. Now they have been replaced by more easily available and longer lasting old tyres. The rosettes of tough, toothed leaves of agaves, and the columnar, spiny stems of some cacti are used as virtually impenetrable living hedges. Efforts are being made to increase the economic exploitation of succulents. Some euphorbias have great potential as a future source of wood and opuntias and agaves are promising economic crops.

WHY SUCCULENTS?

It seems a far cry from the near-desert regions of the world to an average kitchen window-sill, but remarkably the very adaptations which have fitted succulents to their native habitats make them very tolerant house plants. This is because their accummulated storage of water enables them to survive a long time without watering, and they tolerate irregular watering with long pauses, which other plants could not withstand, or at least not without severe damage. Members of the genera *Gasteria, Haworthia, Agave, Sansevieria,* etc., can be seen in homes of people who have not much time for their plants. Succulents could be used even more widely as decorative subjects (the larger types of *Dracaena, Agave, Aloe, Yucca, Nolina, Cereus, Opuntia,* etc.) especially in the interiors of public buildings where they will maintain their striking appearance with minimal care. Succulents are not only undemanding and vigorous, but they have ornamental qualities thanks to their unusual shapes, bizarre growth and remarkable combinations of colours.

On the other hand, there are many other succulents, requiring heated greenhouse conditions, and the care and experience of the specialist grower. These plants are exclusively for collectors. All over the world, there are organizations of cactus lovers, publishing journals on cultivation and the botanical classification of succulents. Why do some people devote their time and energy to growing and propagating bizarre or even inconspicuous plants? It is difficult to discover the motives at work but they are no doubt the same as those which inspire any serious collector or student of the unusual.

SUCCULENTS AND THEIR HABITATS

Succulent plants are found everywhere from tropical regions to within the arctic circle. They are, however, less common in areas where frost occurs in winter months. Frozen water damages their tissues, the water crystals tearing the cells, and apart from the well adapted cacti, only a few species such as *Sedum telephium* and *S. acre,* can withstand frosty winters. The greatest number of succulents grows in subtropical and tropical zones.

Although succulents can occur in unfavourable, dry habitats, in true deserts they are rare. Only the toughest species can grow there and then only in the most favourable places, in hollows, rock crevices, etc., where they can find some moisture. The rains in deserts are rare and irregular. Fog or dew can be a source of water for the plants which have become adapted to it (as in the African coastal deserts, or along the coasts of Chile and Peru in South America), but the true desert is too arid for any plant life, even the best adapted succulents needing some water for survival.

In places which are hot and dry but enjoy a more substantial rainfall, life is more abundant. The rains come every year in the so-called rainy season, and the amount of rainfall is often considerable (200 to 500 mm per year). On the other hand, it is very seasonal, the favourable humid season usually being short, while the dry, hot period lasts for the rest of the year. Because of this regular moisture, the vegetation is more varied, and succulents are abundant. They not only reach the highest concentration of species here, but also the greatest variety of forms. Next to the tall, tree-like cerei, one can encounter the shrubby opuntias, clump-forming agaves and bromeliads, barrel-shaped ferocacti, tiny mammillarias, echeverias, sedums, and the miniature, partly underground cacti. The semi-deserts have many names: the Mexicans call them incorrectly *desierto,* meaning desert, but the botanists call them xerophytic *matorral, isotal,* or *chaparral.*

Many succulents are common in dry deciduous woods with a higher but irregular rainfall (500 to 700 mm per year) where the dry season is shorter and less severe than in a semi-desert. These plant communities include many large, tree-like succulents and even epiphytes (for example Orchideaceae and Bromeliaceae). The ground layer is the habitat of many species suitable for cultivation because they tolerate semi-shade, more regular watering and relatively stable temperatures.

Some succulents are denizens of humid areas. They grow in evergreen forests in tropical regions, and even in the needles underneath the coniferous trees of mountain forests in temperate zones, covered with snow in winter. An interesting group is that of halophytic (salt loving) succulents, growing in saline soil in swamps, lagoons and alongside sea coasts. These plants have special adaptations and many authors do not rank them among succulents (*Salicornia, Sesuvium* and *Batis,* are examples).

It is evident that the term succulent or cactus does not always apply to a desert plant. For the practical needs of cultivation and for information, it can be briefly summarized that the succu-

lent plants come mostly from the warm and sunny regions with low rainfall which occurs only in a few months of the year. The majority of succulents flourish in sunlight and warmth, but they can tolerate considerable fluctuations in temperature. On the other hand, like other plants, they need water for their growth and development, even if they can survive long periods of drought. They also grow at high altitudes above sea level. The genus *Melocactus* is found both on the sandy terraces of the Antillean coast, and in the Peruvian mountainous valleys above 2,000 m. The genera *Sedum, Echeveria* and *Graptopetalum* are encountered in Mexico above 3,000 m as are many of the hardy north American species such as *Opuntia erinacea,* while some *Espeletia* and *Senecio* species grow as high as 4,000 m and over.

BIOLOGICAL ADAPTATIONS IN SUCCULENTS

Loss of water through transpiration is the enemy to plants of hot, dry regions, so the smaller the area of tissue open to the sun, the less the plant will suffer from desiccation. This is why many succulents show such extreme reduction of their surface area, giving them a relatively large volume enabling them to store water without too great a loss.

The water is kept in three basic organs: stems, leaves or roots. We only use the terms stem- or leaf-succulents; the fleshy root formations often occur in either type, but the term root-succulent is not used, and no bulbous or tuberous plants, unless they have some other succulent organ, are classified among succulents. Water is usually stored in special cells, forming the so-called water parenchyma, a transparent water tissue (Fig. 1). It can be found on the surface of leaves *(Trades-*

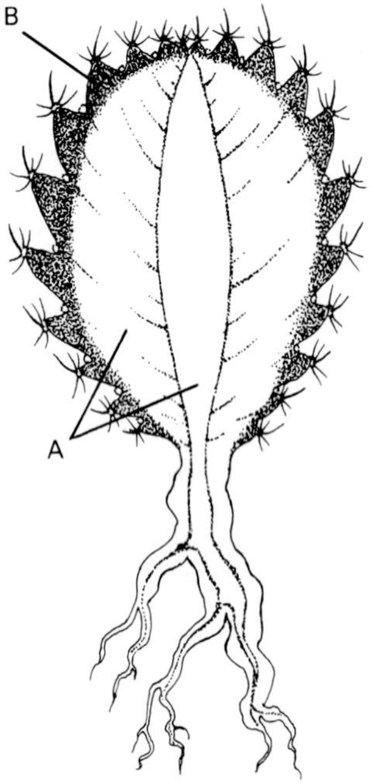

1. In stem-succulents, for example species of *Mammillaria*, water parenchyma (A) is situated inside the stem, chlorophyll (B) is concentrated in the surface layer.

cantia, Hechtia, Peperomia), or in deeper leaf layers (*Aloe*, etc.). In stem-succulents, it is in the inner tissues of swollen stems (*Euphorbia*, and many members of the Cactaceae). It is not pure water but a cellular juice which contains many substances, for example mucins that dry up and protect wounds from infection, or in some cases alkaloids, which in cacti of the genus *Lophophora* have been used as narcotics.

The dimensions of succulent plants differ enormously. There are the tree-like *Aloe*, *Cereus* and *Euphorbia* which can be over 10 m high. *Carnegia gigantea*, *Pachycereus gigas*, and others can weigh several tons. Others are minute in comparison, and their weight can only be given in grams, for example the miniature *Frithia pulchra* or *Blossfeldia minima*.

Leaf-succulents occur in several families, evolutionally considerably distant. There are many forms, the most frequent ones being shrubs, terrestrial rosettes, or rosettes on elongated, often shrub-like branches. The greatest reduction in the number of leaves is reached in some species of the family Aizoaceae, where the leaves are reduced to a pair which are then fused to look one as in *Lithops*. Some species have only slightly succulent leaves (for example *Aeonium tabulaeforme*), while the other extreme is present in the spherical and ovate species of the family Aizoaceae. The leaf-surface form is also very varied, some have smooth leaves, but others are spiny, scaly or tuberculate.

Stem-succulents. It is difficult to draw a line between the leaf- and stem-succulents, because some plants could be included in either category. For instance *Sedum oxypetalum* has a thick, fleshy stem, and in the humid season grows succulent leaves. In the majority of cases, however, we can distinguish the two types. In stem-succulents, the stem has at least partly taken over the leaf's function, that is it contains the green pigment chlorophyll, which other plants have in their leaves. The stem-succulents include many familiar plants, such as the small, free-flowering cacti. Some have segmented stems covered with ribs, tubercles and swellings. In others, the surface is smooth, and the forms are strictly geometrical: ovate, club-shaped and spherical. The species which have spread from unfavourable to favourable conditions, for example to moist forests, are an interesting group. Their reduced stems or leaves become elongated or wider, losing the characteristic features of succulence, and sometimes again resemble their original form, as in the epiphytic cacti *Epiphyllum, Nopalxochia,* and *Rhipsalis.* An interesting phenomenon can be observed by comparing unrelated but similar plants from different continents. For example *Astrophytum asterias* from Mexico (Cactaceae) and *Euphorbia obesa* from Africa (Euphorbiaceae) form almost identical spherical stems. *Senecio stapeliiformis, Stapelia hirsuta, Euphorbia heterochroma, Cereus foetidus,* and *Cissus quadrangularis* form at the first sight hardly distinguishable ribbed stems, although they are not mutually related and originate from different places in Africa and America. These phenomena, called convergence, are frequent and they prove how the environment affects the shape or other characteristic qualities of organisms. Plants growing in similar conditions produce the same shapes and functions regardless of family or geographical location.

Many succulent plants from various families produce **prickles, spines and thorns** (Fig. 2—3). These formations differ in origin: in the family Cactaceae, they have been modified from leaves on lateral side-shoots; in Didiereaceae, they are transformed stipules as in Apocynaceae; in Fouquieraceae, they are remains of petioles when the leaves are shed. In *Euphorbia,* the spines are of diverse origins: within a single genus, they emerge from various original organs. The most characteristic are the spines in cacti, having also the most variable shapes and functions. In shape, they can resemble awls, needles, hooks, bristles and hairs.

The spines can also be differently arranged: some stick out, others are adpressed or clustered. The robust prickly thorns protect the plant from grazing animals, the clear and dense spination shades it from sunlight, and some cacti can absorb water and nutrients through their spines.

Some succulent plants, as protection from the scorching sunshine, heat and drought, remain half hidden underground. The tiny species of the genus *Neochilenia, Turbinicarpus, Fenestraria* and *Lithops,* live almost literally buried all their lives, with only the upper parts of leaves or

2. Many succulents have spines of various types: a—robust, hooked in *Ferocactus;* b—fine bristles in *Rebutia;* c—hanging hairs in *Espostoa;* d—blade-like in *Pediocactus;* e—feather-like; (A—radial spines, B—central spines).

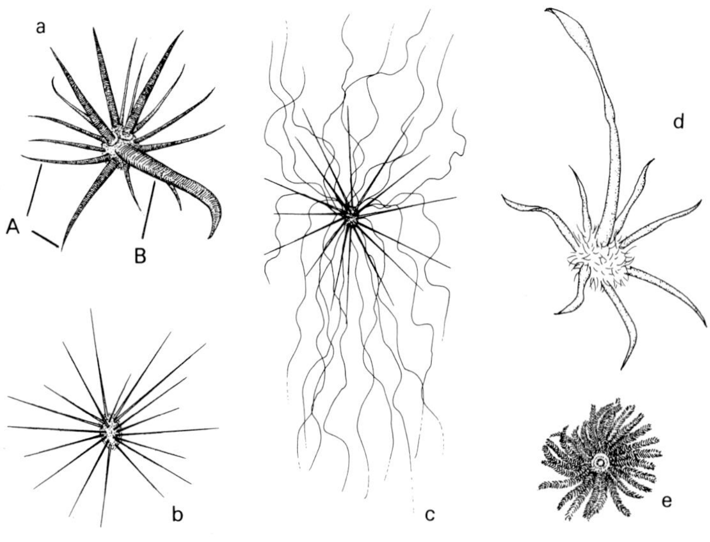

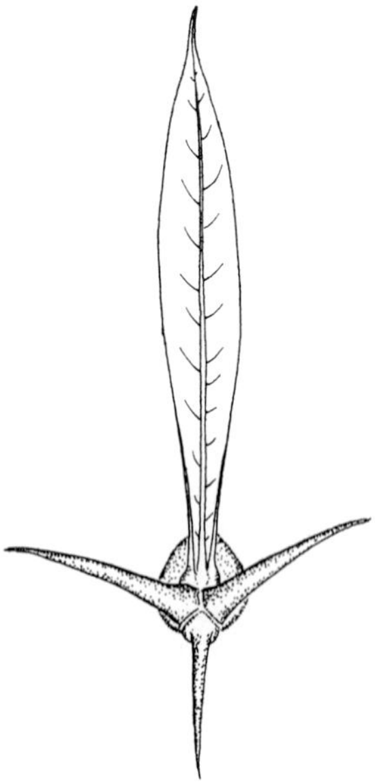

3. In *Pachypodium*, spines originated as part of leaves by modification of stipules. After the leaves are shed, they remain on the stem.

stems at ground-level exposed to the light. There are also species which remain underground only in the dry season. The lack of water makes their organs shrivel, and the contractile roots also play a part. By shrivelling and contracting these roots pull plants underground, reducing the surface exposed to the drought and sun (Fig. 4—5). When water is available, the leaves and stems swell out again and they not only emerge from the ground, but the rosettes open, the concentrated clumps come apart, the tubercles and spines are open at maximum, and like open shutters allow light and air to penetrate to the plant. This means that the plant can make full use of light and water when the living conditions are favourable (Fig. 6). In the dry months, the plants wither and shrink, the tension in the tissues decreases, and as this happens, the spines are drawn closer together rather like slatted shutters and help to protect the plant body during the hottest and driest weather.

The reduction in body-surface is accompanied by a reduction in the number of skin pores, the openings through which water can evaporate from the plant. The pores open only during cool and moist night hours. The epidermis is often strengthened by a thick coating of the so-called hypodermis, massive and solid cells close to the surface. The outer side — epidermis — is covered by a secreted layer of waxy substances (cuticle), and an often colourful frosting of crystalline waxes, is a frequent phenomenon giving rise to blue, grey, and white-frosted succulents. Other species have different forms of covering, hairs, etc.

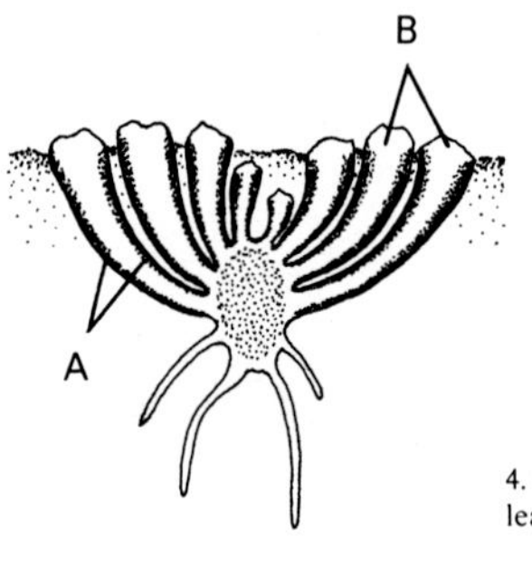

4. Some succulents live underground and the sun rays penetrate through transparent windows (B) into the leaves where chlorophyll is concentrated (A).

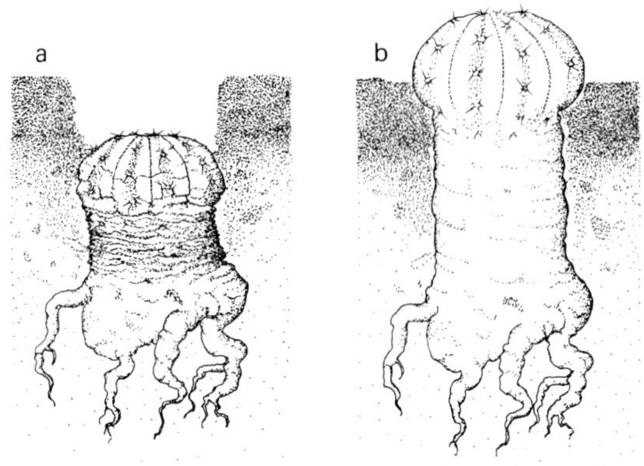

5. Many cacti, e. g. *Echinocereus pulchellus*, remain underground only in the dry season (a), while in the favourable season, the green top is above the ground level (b).

The roots, in the first place, anchor the succulent in the soil, on rocks, and, in the case of epiphytes, on other plants. Their second basic function is to obtain mineral substances and water from the soil. In some plants there is another function, to store and retain water and food reserves. Such plants often produce large, turnip-like or tuberous roots. The root system is often very extensive, spreading over many metres. The roots may grow through the upper layers of the soil, close to the surface.

Many interesting features are found in flower specialization (Fig. 7). Very inconspicuous succulents may produce large flowers of shining and attractive colours. On the other hand, many succulents have reduced the size and number of all the less important parts of a flower, for example many *Euphorbia* species which have no petals or sepals, and replaced them by exuding a large quantity of nectar to attract insects. One has to wait dozens of years for *Agave* flowers, but then the rosette will put all its strength into the formation of hundreds of flowers and thousands of seeds, and subsequently perishes. In the family Asclepiadaceae, the xerophytic and succulent species have developed highly specialized flowers. The same applies to bromeliads, orchids, and some cacti, where the pollination can be carried out by a particular animal only.

6. The spines often function as a shutter, for example in *Mammillaria*. In the dry season (a), tubercles are closed and spines adpressed; on absorbing water (b), the plants open and allow sunlight to reach the epidermis.

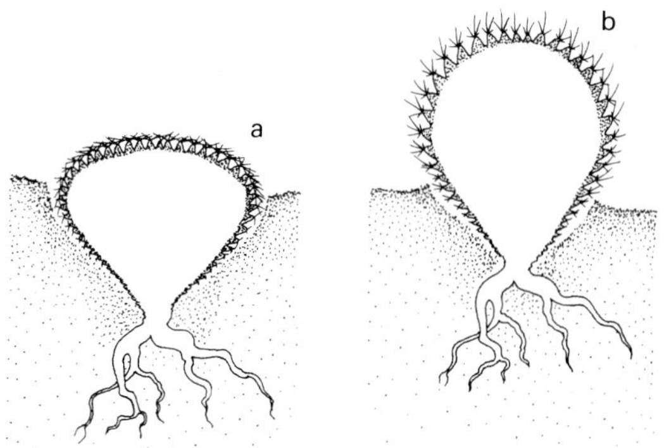

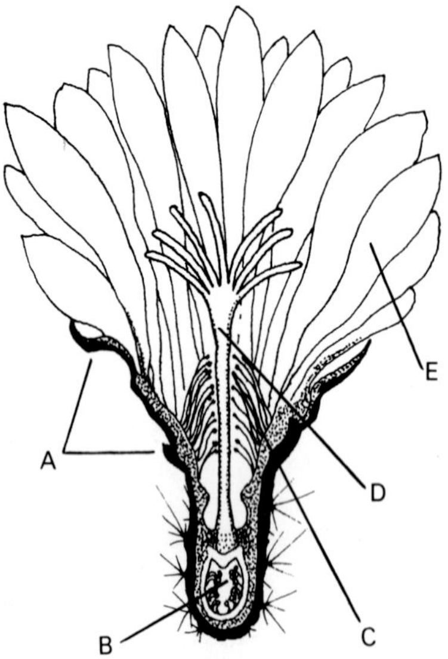

7. Flower structure in *Notocactus:* A — outer perianth; B — ovary; C — anthers; D — stigma; E — inner perianth.

There are succulents which produce fruits without opening flowers at all, this is called cleistogamy.

The fruits and seeds are of diverse shapes. In many plants, the seeds are distributed by the wind. These species have seeds equipped with flying devices such as hairs or membranous wings, or they are especially light and flat. Another large group depends on distribution by water in rain and floods, the seeds being bigger and capable of floating. Other species depend on animals, mainly birds, eating the juicy fruits and spreading the seeds in their droppings (for example in *Opuntia, Mammillaria, Cereus, Cissus,* some *Aloe, Bursera,* etc.), or the minute seeds may get stuck to the feathers and feet of the birds (epiphytic *Rhipsalis,* etc.). The fruits of *Euphorbia* split at maturity, and shoot seeds to a distance of several metres.

Studies dealing with the water economy in succulents have discovered many physiological adaptations. The most important discovery was the so-called Crassulean Acid Metabolism (CAM), found in the majority of succulent plants. This allows these plants to carry out photosynthesis, even with closed pores. They can capture and store carbon dioxide at night, when the air is cooler and moister, and when they are able to open their pores. Many succulents contain toxic or foul-tasting alkaloids and latexes protecting them from pests and parasites. The seeds of many desert succulents will germinate only in the season when conditions are favourable so ensuring that some will start to grow when the weather in their homeland is propitious.

BOTANICAL CLASSIFICATION AND NAMES OF SUCCULENTS

Succulents, like all other plants, are subject to the rules of taxonomy and nomenclature which must be respected and followed since they are the means through which specialists, collectors and amateurs can understand each other.

The *species* is the basic term of botanical classification. It is a group of similar and related individuals, producing identical offspring and not naturally forming hybrids with other species. Within a species, lesser divisions are recognized; for example *subspecies,* while below this is the *variety.* The term subspecies is used to denote the differences in some geographically separated natural populations which as a consequence of further evolution may in time give rise to new, independent species. They are designated by the abbreviation *ssp.* which is placed before the subspecies name *(Sempervivum montanum* ssp. *carpaticum).* Variety is used to describe the less substantial differences in morphology between various populations of the same species such as hairiness, size or spine length, but the differences must still be hereditary. For these the abbreviation *var.* is used before the varietal name *(Thelocactus bicolor* var. *bicolor).* If the different features within the plants of one species are manifest only in individuals of one population in one site and they are not permanent, hereditary characteristics, then the plants are denoted *forma* with the abbreviation *f. (Gymnocalycium mihanovichii* f. *rubra).* Above the species comes the *genus* (plural *genera*). It comprises various related species which have some important common features, for example the structure of the flower, fruit, seed etc. Two lesser divisions are sometimes used within the genus and are particularly useful when discussing cacti. These are the *series* and *subgenus.* They are used as a matter of convenience particularly when dealing with a genus which contains a large number of species. The series is the lower ranking, while in turn a number of series can be grouped together as a subgenus, abbreviated as *sg.* or introduced in brackets after the generic name (e. g. *Mammillaria (Dolichothele) longimamma,* etc.).

Plants have two names, for example *Thelocactus bicolor.* The first name indicates the genus, while the second is the specific epithet. If several varieties are distinguished in a species, their names follow after the abbreviation *var.,* for example *Thelocactus bicolor* var. *tricolor.* The name is usually followed (but it is not a rule) by the name of the author who first described the plant (or the recognized abbreviation of his name). Where the plant has been reclassified (recombined) into another genus, or later authors have re-evaluated its status (if the plant was transferred from the rank of variety to a species for example), the name of the original author is in brackets, and the author of the accepted combination is at the end. For instance the original *Cereus thurberi* Engelm. was later put into another genus, and it exists under several names: *Pilocereus thurberi* (Engelm.) Rümpl., *Lemaireocereus thurberi* (Engelm.) Britt. et Rose, or *Stenocereus thurberi* (Engelm). Buxb. This demonstrates that a single species can be known under several names, called synonyms. One can also encounter only the generic name, for example *Cereus* sp. where the author indicates a yet unspecified or unknown species, or uses the plural of the word species by its abbreviation *spp.* (not to be confused with *ssp.*).

Beside the plants known in the wild (the so-called botanical species), there are also taxons which have originated from selection or hybridization with man's participation. They are called *cultivars,* and receive names usually of a descriptive or commemorative sort, and not of Latin form. They are designated by the abbreviation *cv.* after the name, or by single inverted commas and roman type as in *Crassula* 'Morgan's Beauty'.

The systematic arrangement of plants into species and genera is not simple and it is far from being finished, mainly because experts disagree as to what makes a true species. Some taxonomists believe in a broad concept of a species and include under one name entities which others would keep separate. Others prefer to split up the same entities into a number of species holding that small differences are enough to justify the division. These two schools of thought are familiarly known as lumpers and splitters. This disagreement over what constitutes a species has had rather drastic effects as far as the naming of succulents is concerned, cacti in particular being incredibly variable, responding quickly to a change in their environment and looking quite different under different conditions. This is particularly noticeable with plants in the wild and in cultivation.

To give an indication of the confusion which can arise, C. Backeberg in the early sixties described in his monograph on the Cactaceae 220 genera and 2,700 species. In the seventies, D. R. Hunt recognized only 120 genera and 2,100 species while W. Barthlott has found that altogether about 14,000 different names have been used over the years. Even now, every year dozens of so-called new species are described in the literature, many of them barely distinguishable from species already known and hardly meriting specific status.

Taxonomy, that is botanical classification, will however always remain in the hands of the experts and depend upon their subjective evaluation. Growers and collectors have to accept this fact and be prepared to find their plants given different names by different authors, making up their own minds as to which to use for themselves.

GROWING SUCCULENTS

When growing succulents in European or North American conditions, we have to respect the fact that almost one half of the year is unfavourable for their growth. The shortage of sunlight and warmth forces us to keep the succulents in dormancy over winter. At that time, they require minimal attention: they should be placed in a sunny spot at a temperature of about 7° C, and they need no watering for several months. Others, particularly the leaf succulents, are more demanding and less tolerant of poor winter light. They require careful and sparing watering, and temperatures need to be higher. For a majority of succulent species, spring, summer and early autumn are the growth seasons. Some south African species end their dormancy in late summer, growing during autumn and winter. This complicates cultivation, because the plants have to be watered during the period when light intensity is at its lowest.

POSITIONING

The most important requirement for the successful cultivation of succulents is a location in a well-lit place, reached by the sun's rays for at least half each day. In the home it means a position near the window, if possible on the windowsill (Fig. 8). Only a few species can be grown farther from the window (for example *Hoya, Sansevieria* and *Yucca*) but they prefer good light. Rooms facing north are better suited for other plants. The large-sized *Agave, Yucca, Nolina, Opuntia* and *Cereus,* used on staircases, in halls, etc., should be placed in summer in the open, where the weakened, light-starved shoots become visibly better.

Most succulents thrive in greenhouses, where the plants enjoy more light, warmth and humid air, and these conditions can be more easily regulated. Moreover, the plants can be over-wintered in a well equipped greenhouse, and this sort of cultivation is best in temperate conditions. The greenhouse for desert succulents has to be situated in a sunny site, protected from the wind. It should be sited so the long axis runs from east to west to make the most of all the available sunlight. Custom-built houses of either aluminium or timber (western red cedar is far the best here) are very good for growing cacti, though extra ventilation is usually an advantage. In any case, installation of heating is recommended, for example by a connection with central heating, or independent heating by oil, gas or electricity. This prevents damage from sudden drops in temperature. It is most economical to site the heat sources under the benches to use maximum energy for heating the plants and not the roof. Ventilation is also important, because on sunny days, the plants can be overheated and scorched. This risk of scorching is reduced when a fan is installed, maintaining the air circulation in the greenhouse.

TEMPERATURE

Most succulents require high temperatures for their optimal growth, tolerating temperatures above 40° C without damage. During the growing season the temperature should not fall below 18° C. The tropical species in particular thrive in warm conditions, but the plants from higher mountain elevations and regions farther from the equator prefer the lower limit of optimal temperatures. It was proved recently that the species adapted to life in the most unfavourable desert habitats require varying temperatures for successful growth, that is they do best if the heat during the days is replaced by a strong nocturnal chill. This applies to some mountain succulents too (for example *Tephrocactus, Lobivia,* etc.). We have to realize that the temperature is one of the factors affecting the plants, and to keep the plants at their best, their temperature requirements must be respected. For instance, a high temperature with high humidity produces growth, as does a high temperature and shade, but high temperature, drought and strong sunlight stop the growth.

LIGHT

In the life of plants, light is responsible for most of their processes — it is the source of energy, affecting the shape of organs, periodicity of growth and reproduction. In poor light, growth slows down, the plants are deformed, starved, and flower poorly, their typical colouration changes, they become less disease-resistant, etc. Many succulents in the wild have become adapted to intense light, and they show modifications to help them survive. In poorer light the spines and hairs are less robust and dense, the stems are longer, etc. Changes in light-intensity are particularly dangerous — the bright spring sunshine coming through glass can in some regions scorch the plants, which have to be accustomed gradually to the sun rays at the beginning of their growing season by shading either with tinted glass or some form of blinds. This is not a problem in northerly climates.

8. In interiors, a window frame is the best habitat for succulents.

WATERING

Except in some rare cases, succulents absorb water by their roots. Water contains minute amounts of a number of minerals which help to provide the plants with necessary nutrients. The roots will, however, only take in these nutrients and pass them on when they are in a healthy condition. The water should ideally be soft, preferably rainwater, though this is by no means essential. For tropical species it is recommended that its temperature is higher than that of the air.

Many species from arid regions are sensitive to excessive water in the soil, especially when the temperature drops, and it remains cold for some time. On the other hand, they respond well to irregular watering — a period of liberal watering in warm weather followed by a thorough drying of the soil and roots lasting several days. Superfluous water in the soil together with cold stop the activity of the roots and encourage the growth of fungi and bacteria, making the roots rot and sometimes the plant perish. The correct application of water at the right moment is one of the most important factors in growing succulents. The requirements of various species can be guessed at by looking at their root system — the species with turnip-like or tuberous roots are usually more xerophilous. There is no exact recipe as to when and how much water should be applied to plants, but it is possible to learn the right moment. That requires close observation of the plants, practice and a willingness to learn from one's failures.

SOIL

Bearing in mind the soil conditions in the wild, it is essential that any rooting medium used in cultivation should be well aerated and free draining. Many fancy potting composts have been, and still are, devised by enthusiastic growers. There is, however, plenty of evidence to show that any of the standard potting soils are suitable provided watering is done carefully. Ideally a loam-based mixture such as the John Innes potting compost No 1. is ideal with extra grit or coarse sand added for the more exacting species. Even the all peat mixes have been used with success but are best blended with $1/2 - 1/3$ part of coarse sand. Once a plant has been in its pot for a year, supplementary feeding will be essential. This can take several forms. Slow growing species are best top-dressed in spring, removing the top layer of old compost and replacing it with fresh. Alternatively a dry, general fertilizer can be sprinkled around each plant in spring. Fast growing plants are best fed with liquid fertilizer at 14 day intervals during the growing season. A high potash feed should be chosen such as that formulated for tomatoes.

REPOTTING

The best time for repotting is at the beginning of the growing season, that is usually in spring. If it is necessary, it can be undertaken at other times too, but it is not recommended in winter, when damaging roots can mean the death of the plant. During repotting, all the compost must be changed, even in the root-ball, from which the soil should be shaken; roots are shortened to one third of the original length. The sensitive species can be left dry for a few days before repotting, to allow any damaged root surfaces to heal.

The plant is replaced in the pot and the pot then filled, preferably with an absolutely dry compost, and the plants are never watered immediately. Tall specimens are best tied to a cane support until their roots have become established again. The repotted plants are put in a warm, shaded, humid place. Careful watering is applied only after several days, enough to prevent the root system from drying up completely. Only when the plant begins to grow again, when the first buds appear or the tops take on a richer colour, should more liberal watering be allowed.

Some succulents can be planted or plunged outdoors for the summer. Early in October, they have to be taken back inside, to protect them from frost. If they have been planted out of their pots, when dug up they can be put in wooden boxes close to each other, and the roots buried under dry peat or peat with sand. No watering is necessary during winter, especially for larger specimens, which can survive better in dry conditions. Plunged plants are simply returned to the greenhouse bench.

NUTRIENTS – FERTILIZERS

Providing that light, temperature and moisture are optimal, the speed of growth as well as the colour of leaves, length and thickness of spines and number and size of flowers is determined by nutrients. An application of fertilizer can delay the necessity of repotting. Where the succulents are cultivated in a sterile substrate, the nutrients have to be regularly supplied during the growing season.

Fertilizers are applied only when the plants are growing, that is in spring and summer, and they have to contain balanced amounts of nitrogen, phosphorus and potassium. Various suitable products are available in garden shops. A high potash fertilizer sold for tomato growing is very satisfactory.

PROPAGATION

Succulents are propagated in the same way as other plants; there are two possibilities – generative propagation, that is from seed, and vegetative propagation, that is from cuttings, offsets, suckers or by grafting.

Both ways have their advantages. From seed, we can obtain genetically different individuals which can show differences within the natural variability of the species. The more vigorous or attractive specimens can be selected from them. By vegetative propagation, we can obtain offspring identical with the parent plant. Individual plants can yield seeds only if the species is self-fertile. Vegetative propagation is essential in the reproduction of various mutations: colour mutations, cristate and monstrous forms, and in the propagation of cultivars.

Propagation from seed is one of the most interesting facets of the cultivation of succulents, and provides much useful information. The first prerequisite is to obtain the seeds. Commoner species can be bought from garden shops and centres, less common ones can be obtained from specialist societies, or from home grown plants. When they are flowering, the pollen from the anthers of one plant has to be transferred, usually with a fine brush, to the stigma of another specimen of the same species. Only some species are autogamous (self-fertile), that is the pollination can be carried out within one flower. Most succulents are allogamous (self-sterile), and at least two plants are needed in flower at the same time. The seeds of many species germinate immediately after collection, but the best season for sowing is in spring, and the seeds have to be stored in a cool, dry, dark place. In late spring or early summer, the temperature in greenhouse or on a window-sill is suitable (20–27° C) for germination. Sowing can be done in other months, but it then requires additional heating or lighting. The seeds are sown in plastic containers or trays with transparent or translucent lids, or in pots enclosed in plastic bags. The containers should be washed in disinfectant or at least in hot water. They are filled with the sowing mix, for example with a mixture of equal parts of sand, peat and quality, sterilized loam or a standard mix such as John Innes Seed Compost. If sterilized soil is not available, it is advisable to water the sown seeds with a fungicide such as Captan or Benlate. These chemicals are also used when fungi attack the germinating seeds. The ideal temperatures for the germination of tropical species are about 25° C, in species from cooler areas about 20° C. High humidity and light are important factors, but sunlight is dangerous, and the trays should be kept in shade. After several weeks or months, the seedlings are pricked out into larger trays. This is done with a special dibber or a pencil, which is used to make holes for the roots. The seedlings are fast growers and therefore they demand nutrients. A diluted solution of high quality tomato fertilizer can be added. It is desirable that the seedlings get as big as possible in the first season. The optimal temperature of 20–30° C and a moist compost must be maintained, and feed is applied. In these extremely favourable conditions seedlings may turn soft and lack vigour and it is recommended that they are hardened off gradually. It is best to leave the seedlings to dry off after two to four weeks of intense growth. They are not watered for several days, and will begin to wither and shrivel a little. Their reaction to the next watering is surprising: they soak up and grow fast.

Vegetative propagation uses the regenerative capacities of the plants (Fig. 9). The cuttings, offsets and scions are cut off with a sharp knife and the base of the cutting is covered with

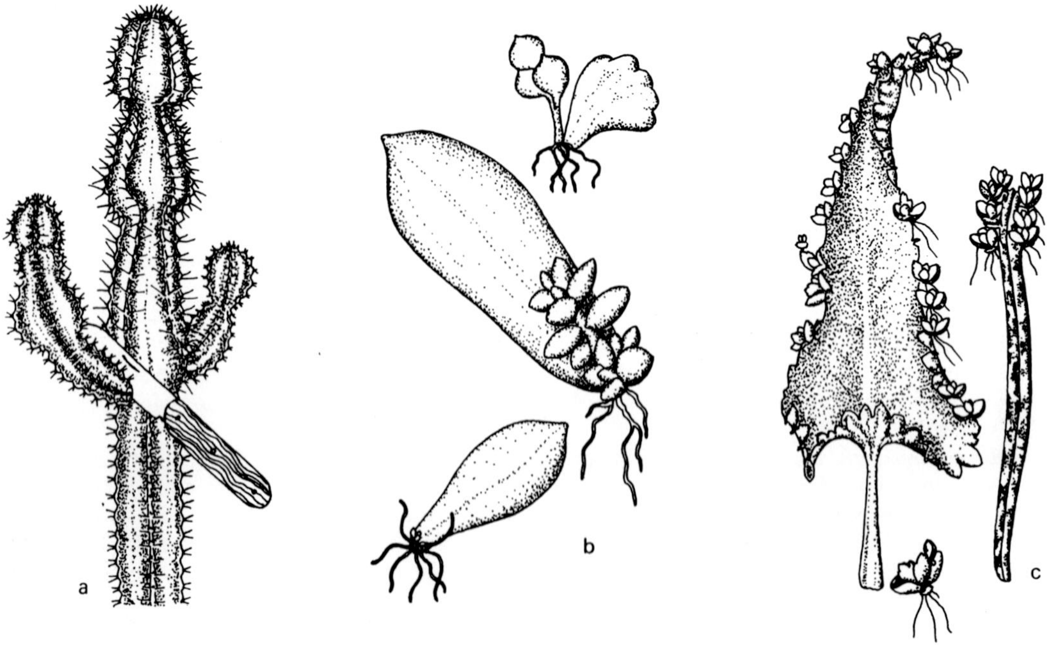

9. Stem cuttings are cut off with a sharp knife at the narrowest place and left to callus for a few days in the shade (a). In some succulents (mainly Crassulaceae), leaves can be detached and put to root on wet sand (b). In *Kalanchoe*, young plants grow on leaves which are still on the plant (c).

powdered charcoal or hormone rooting powders. The stem-cuttings of leaf-succulents need their lower leaves removed and are best left for several days to allow the cut surface to dry off. They are then inserted into a mixture of peat and sand and covered with plastic sheeting or glass. The compost is kept slightly moist. Single leaves of many members of the Crassulaceae can be taken as cuttings, simply laying them on moist sand until leaves and shoots arise.

For the species which are difficult to root, which grow slowly, or are special clones, grafting is the only method (Fig. 10). Fast-growing succulents are used as stock. In cacti, the best stocks are provided by cerei: *Cereus peruvianus, Eriocereus* spp., *Hylocereus* spp. and *Selenicereus* spp. Small seedlings can be grafted on *Echinopsis* spp. or *Pereskiopsis.* Other families of succulents also have a wide range of stock materials: Apocynaceae can be grafted on oleanders; Asclepiadaceae on the tubers of *Ceropegia woodii,* or on the stronger growing species of *Stapelia* and *Ceropegia;* Didiereaceae on the vigorous *Alluaudia procera;* Crassulaceae (the species with elongated stems) on *Kalanchoe tubiflora* and *K. daigremontiana.* Since grafting is really a surgical operation, it requires a strictly clean environment. The two tissues can be united only when the plants are in growth, so we carry out the grafting in the warm months of the growing season. Dull and damp days are not recommended, because there is an increased danger of infection. It is important to ensure that the two surfaces are united, without any air bubbles, and the vessels of the stock and scion have to fit together. Slipping is prevented by rubber bands, spines, or light weights.

PESTS AND DISEASES

The succulent plants in cultivation are vulnerable to attacks from specific pests, rare in other plants. Of the sucking parasites, root mealy bugs *(Rhizoecus falcifer)* and mealy bugs *(Pseudococcus maritimus* and *P. citri)* often occur. A related species of these pests is the scale insect

Dactylopius cacti, which was used as material for the famous red dye, cochineal. This insect was grown on opuntias for this purpose. In cultivation, the pseudococci are the most serious nuisance — they thrive in a warm and dry environment, and have ideal conditions for reproduction in succulent greenhouses. Their reproduction cycle takes three to four weeks only. The root mealy bugs live in the soil on roots and around the collar of the plant, especially places where they are not affected by water, concentrating on the crown, beetween the ribs and tubercles, forming white woolly clusters. Another common pest, *Diaspis cacti,* is visible near the ribs on leaves and stems in the form of light, circular dots. Aphids can occasionally become a nuisance, especially on leaf-succulents. The above mentioned sucking insects must be completely eliminated from the collections, because they often carry fungal diseases, virus and bacterial infections, and weaken the plants. Treatment is never easy and requires poisonous chemical sprays, now available on the market. The treatment usually has to be repeated. The mites *Tetranychus* (red spiders), scarcely visible to the naked eye, are a dangerous group able visibly to destroy the epidermis around the crown in a short time. Insecticides which control these pests must always be used with care and the instructions followed to the letter.

The most serious pests living on roots are the nematodes (eel worms). They form tumors and knots on the roots and prevent them from functioning. They are difficult to remove. The most efficacious method is to dig out the plant and soak it in hot water above 54° C. A ten-minute bath will kill almost any pest. The temperature must never exceed 55—56° C, and the method can be applied only in cacti, euphorbias and some other resistant species. The leaf-succulents and epyphytic cacti will not tolerate it.

Some other pests are more rarely encountered: for example the tiny dark flies *Sciara,* whose larvae attack seedlings. Various larvae, grubs and caterpillars devouring seedlings, roots or adult plants are sometimes found in the soil. In dark corners, one can at night find earwigs and other insects, eating the roots or green parts of plants. All these pests are destroyed by poisonous

10. When grafting on thin and long stocks, as in *Pereskiopsis, Pereskia, Allaudia,* etc., the position of the scion and stock is best fixed with a spine (a). When grafting on stronger and shorter stocks, the connection of vessels (v) is important, and the stock and scion are best held with a rubber band (b). Seedlings can be grafted onto the stock (c).

sprays, or the affected plants are watered with poisonous solutions. In humid greenhouses or in outdoor cultivation, the plants suffer attacks from slugs and snails, which can be trapped by poisoned bait available in pellet or spray forms.

Dangerous and unfortunately frequent diseases are initiated by the lower fungi (moulds). The infestation is caused by the species of the genera *Botrytis, Fusarium, Helminthosporium,* etc. They bring about rotting (especially in winter), toppling of seedlings, dark necrotic marks on the leaves, destruction of roots, etc. The fungi of the genus *Fusarium* can live in the tissues for years and are the cause of growth defects, deformations and browning. The treatment of fungal diseases is carried out chemically, but they are more acute if the living environment of succulents deteriorates: for example in cool and humid summers, in winter or in poorly aerated and over-shady greenhouses. Sunshine, fresh air and dry places are enemies of fungi. In case of an infection, watering is stopped, the plants are situated in fresh air and sunlight, and a chemical preparation can be used. The basic rule for the use of chemicals is to apply them in the evening of an overcast day, and to aerate the greenhouse thoroughly on the following day. At higher temperatures and in the sun, many chemicals are dangerous for plants. Precautionary measures may also have to be taken because some substances are harmful to man.

The infections caused by virus or bacteria are dangerous, but since their cure is practically impossible, the only action is prevention of a further spread of the trouble. Remove and destroy all infected plants immediately from the collection.

The physiological disorders include the consequences suffered from lack of light — pale, yellowish or whitish shoots, overelongated new growth, etc. The plants have to be transferred to a better lit position. In the warmth-loving species, a cool environment will bring about dark patches or black tips of leaves. The requirements of the plants have to be respected. An unbalanced supply of nutrients, or a shortage in some substances can produce deformations, yellowing or whitening of young growth. Repotting may be the answer. Loss of roots by rotting is usually the result of overabundant watering, especially at lower temperatures; the affected parts have to be cut off and the plant allowed to grow new roots.

1

Chapter 1 CACTI OF NORTH AND CENTRAL AMERICA

The cactus family (Cactaceae) covers leafy plants (Pereskeideae), stem-succulents with reduced leaves (Opuntioideae), or with entirely absent foliage (Cactoideae). The stems, also called bodies, are either segmented or ribbed with a tuberculate surface structure. The family is characterized by areolas (spine-cushions); these are in fact greatly reduced side shoots (brachyblasts). The spines are actually modified leaves: awl-shaped, needle-like, hooked or straight, often in the form of short or long hairs and feathery prominences. The dense, light-coloured spines protect the plants from sunlight and evaporation; the robust, prickly needle-like and horny spines deter herbivorous animals. The spines are thought to be capable of absorbing moisture from fog or dew and supplying it to the plant.

Cacti are confined exclusively to the American continent, from Canada to Tierra del Fuego, ranging from coastal regions to elevations of 4,000 m. The genus *Rhipsalis* is the only exception, several of its species growing in Madagascar, east Africa and Ceylon (Srí Lanka). The largest, tree-sized genera *Lemaireocereus, Pachycereus* and *Carnegiea* can reach the height of 10 to 14 m and several tons in weight. On the other hand, the smallest species, *Turbinicarpus klinkerianus,* grows up to 1 cm in the adult state and weighs only a few grams. Cactus flowers have striking reddish, yellow and white colours, but bluish and greenish shades also occur. They are either night-blooming, opening for one night only, or open by day and are longer lasting.

Vast areas of North America and the adjacent Central America have a dry climate; they are covered by deserts, semi-deserts, or by dry deciduous forests *(chaparrals)* or by bushy woods *(matorrals).* They provide habitats for a variety of cacti; for example great numbers of *Neobuxbaumia (Mitrocereus) macrocephala* (Web.) Dawson (1). The cacti often determine the shape of the landscape and dominate other plants.

Ariocarpus fissuratus (Engelm.) Sch. (2, 4) is a small flattened plant with a large turnip-like, water-storing taproot, growing literally buried underground. The top of the plant body, situated at ground level, is divided into tubercles, which are fissured and wrinkled. The surface layer (epidermis) is grey-green, sometimes reddish, and in the wild is always covered by clay and dust, which protects it from sunshine and makes the plant well camouflaged.

2

3

This imitation of the environment is called mimicry. Only in the flowering stage is the cactus very noticeable with its conspicuous purplish-pink flowers. The tissues contain bitter and toxic alkaloids, especially hordenine, protecting the plant from herbivores as efficiently as the spines do in other cacti. A group of related species, *A. agavoides, A, scapharostrus, A. kotschoubeyanus* (3) live in a similar way in alluvial sediments of secondary rocks rich in calcium. The flowers of these species are coloured by the purplish-red pigment betalaine (phytocyanine), typical of the family Cactaceae and several other families of succulent plants.

All the miniature 'underground' members of the genus *Ariocarpus* are rare plants in great demand and difficult to cultivate, particularly because their roots are very sensitive to over-wet soil and to the presence of decaying organic matter. Bacteria and moulds rapidly infect and destroy the whole bodies. For success they require a porous soil with 30 to 50 per cent of sand or a coarse loam-based mix, and watering only in warm summer months following a thorough drying out of the compost. In winter, an absolutely dry environment and a temperature of about 12° C are necessary. Propagation is by seeds. Although it is very demanding and requires patience and a sensitive approach, it is the only way of preserving these plants in cultivation. Seedlings can be grafted onto the stock of *Echinopsis* cv. and *Eriocereus jusbertii.* They grow fast and produce flowers quickly and losses are minimal.

Ariocarpus kotschoubeyanus (Lem.) Sch. (3) is the smallest member of the genus. It frequents fine, flat clay-sandy sediments or gravel mixed with clayey sand. The

4

5

tuberculate rosettes measure 5 to 7 cm. A white-flowered variety occurs in Mexico in Tamaulipas, while the varieties from central Mexico, from the states of San Luis Potosí and Querétaro, have carmine blossoms.

Ariocarpus furfuraceus (Wats.) Thoms. (5) is a robust species up to 25 cm in diameter with keel-shaped, triangular tubercles and white to pinkish flowers. It is found on crumbling limestone hills, in full sunshine. The turnip-like root has a network of ducts filled with water-retaining mucus serving as a water storage. *A. retusus* is closely related, or regarded as a variety of the same species. It grows throughout an extensive area of the Chihuahua Desert.

Ariocarpus trigonus (Web.) Sch. (6) has a flattened, bright green body composed of numerous tubercles. Unlike the preceding species, the epidermis has a fresh green colouration and the flowers are yellow, in larger specimens growing around the outer edge of the top of the plant body. In the wild, this cactus grows under shrubs protected from the sun, but in cultivation, it also requires warmth and careful watering.

The genus *Astrophytum* — as the name implies — represents cacti more resembling starfish than plants. Their bodies are tough, the epidermis is often covered with flakes of white wool and the ribs are pronounced. All the species (about six) have yellow, abundant flowers. They are attractive cacti in great demand.

Astrophytum capricorne (Dietr.) Britt. et Rose (7). Similarly to the closely related *A. senile,* even small plants are capable of flowering; the flowers are red throated, the spines are long, soft and bent upward. Both species come from northern Mexico, growing there on limestone together with succulents of rosette form and bushes *(matorrals),* either exposed to the sun or in semi-shade, protected by vegetation. Several varieties of *A. capricorne* are distinguished, differing in the shaping of ribs, setting of spines and colour of flowers. *A. capricorne* var. *niveum* has the most beautiful and delicate flowers; *A. c.* var. *crassispinum* has yellow flowers which lack the red centre.

Astrophytum ornatum (DC) Web. (8). A slender plant at the right side of the picture. It is native to central Mexico, found in the deep rifts of the river Río Montezuma. The name *ornatum* (decorative) is justified by the arrangement of golden-yellow spines, tall ribs partly covered with flakes, and by the slender stem. The flowers do not occur on young plants. *A. ornatum* is less often seen in collections than other astrophytums; it is scarce in the wild too. The opportunity of obtaining seeds is limited, and like all the species, two plants must be cross-pollinated to obtain seed.

Astrophytum asterias (Zucc.) Lem. (9) is the most bizarre species of the genus, the flat ribs with scattered flakes and felted areoles resembling the skeletons of sea-urchins. In its mode of growth it is more like *Ariocarpus:* it remains

6

7

8

half-hidden underground, with only the tip of the crown emerging, hardly distinguished from its surroundings. The centre of its distribution is in eastern and northern Mexico, ranging to the south of Texas. It frequents scrubland on gravel sediments with fine clay. In summer, it can become overgrown by grass. The flowers are produced throughout summer: they are yellow with a red throat. Apart from *A. coahuilense,* it is the least straightforward species to grow.

Astrophytum myriostigma var. **nudum** (R. Mey.) Backb. (10) and var. **potosinum** (Möll.) Krzg. (11) are popularly known as the Bishop's Cap because of the unique shape and structure of their stems. This species has its ancestral home in the central part of the Chihuahua Desert. The large area over which it is distributed contains variable populations, differing in the number of ribs, size of flowers, presence or absence of felted flakes. Flakes are absent for example in the var. *nudum,* as implied by the name. This cactus can be easily propagated from seeds. It can be hybridized without difficulty. The numerous hybrids are not valuable from the collector's point of view, but possess striking, bizarre combinations of features inherited from the parents.

Aztekium ritteri Böd. (12) is the only species in its genus, one of the most interesting miniature cacti from Mexico. Since its discovery in 1928, it has been greatly sought by collectors, despite the rather unsuccessful attempts at cultivation. Until recently, the demand was covered by imports from its only known location in Nuevo León. Propagation in the home is made difficult by its fine seeds and the slow growth of the seedlings, which average less than 1 mm in diameter after several weeks of growing. Grafting these miniature specimens is very difficult, although some patient cactophiles can graft them on *Pereskiopsis* spp. or *Echinopsis* cv. The seedlings then rapidly increase in size and the first flowers appear within 2 to 3 years. The seedlings can further be propagated vegetatively, by grafting. The cactus must be kept in a warm and sunny

9

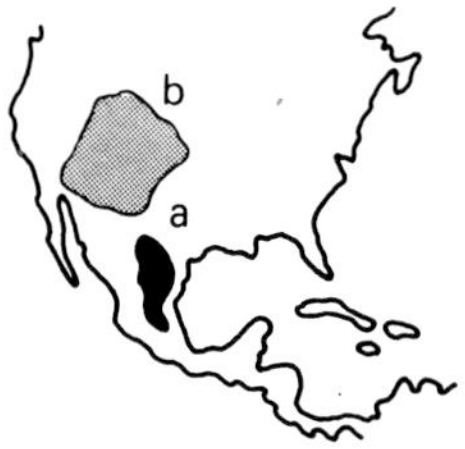

Distribution of the genera *Astrophytum* (a) and *Sclerocactus* (b)

10

11

place in a greenhouse, and watered sparsely in warm weather. The compost should be free of organic matter. The flowers appear throughout the summer season; they are pale pink to white. In nature, *Aztekium* frequents steep slopes and gypsum rocks. Hundreds of adult and thousands of young specimens cover the walls of rifts in places untouched by the sun. The rock is crumbly and weathers rapidly, therefore most of the sites are inaccessible to man so the cactus is thus so far safe from man in its habitat.

Some 60 species of the genus *Coryphantha* grow in arid semi-deserts and grasslands as well as in more humid deciduous woods.

Coryphantha elephantidens Lem. (13) is up to 25 cm high with a deep green body and conspicuous pale spines. The carmine to white flowers, always with red throat, appear in the second half of summer. *C. elephantidens* comes from the Mexican state of Morelos. The similar, yellow-flowered *C. bumamma* ranges to the south and west. Both species grow in black soil rich in humus, on limestone in lush grass — in places with profuse summer rainfall. They are easy to cultivate, requiring a great deal of nutrients and watering in summer, in winter surviving in a completely dry environment at about 10° C. They are suitable for beginners and growers with modest conditions.

12

13

14

Coryphantha macromeris Engelm. (14) is a soft-fleshed plant with prominent spines and grows in clumps. It frequents dry situations and it is sensitive to watering. The large hairy flowers appear at the end of summer and last only for 1 to 2 days. It occurs in the south of the USA, in western Texas and mainly in northern Mexico.

Coryphantha calipensis Bravo (15) is found to the south-west of the town of Tehuacán in Puebla, Mexico. It grows in small clumps; the tubercles are elongated and curved upward. The flowers are pale yellow. The crown, areoles and the furrows on the tubercles are in cultivated specimens usually enveloped in white wool.

Coryphantha reduncispina Böd. (16) is an inhabitant of flatland pastures in the Río Soto La Marina watershed in Mexico. The yellow flowers and the body structure reveal a close relation to *C. cornifera* from central Mexico. Cultivation is as for *C. elephantidens.*

Echinomastus unguispinus var. **laui** (Frank) Glass et Foster (17). This genus embraces some 10 species distributed mainly in the American part of the Chihuahua Desert. They are exacting cacti, difficult to cultivate. Varieties *laui* and *unguispinus* are relatively the easiest to grow and come from central Mexico. Adult plants require a loam-based compost and irregular, sparingly applied watering, given only on hot summer days. The flowering season is early in spring, in April and May; the flowers are green-brown.

Echinomastus warnockii (Benson) Glass et Foster (18). The species of *Echinomastus,* now sometimes classified under *Thelocactus,* are notable for their strong spines which make a starry mesh all over the plant body, only pierced by the flower buds which open in a cluster. This species can be propagated successfully from seeds. Growth is, however, slow and some growers prefer to graft, though the plant is quite successful on its own roots. The specimen in the picture is a three year old seedling, flowering freely several times in summer, for two seasons. A maximum of sunshine and warmth is a must not just for flowering, but for the survival of the cactus.

Cephalocereus senilis (Haw.) Pfeiff. (19, 20). Few cacti are so well known as this Old Man Cactus. The ten metre high bodies enveloped in silvery hairs stand erect along the limestone and gypsum valleys in the state of Hidalgo. These giants are of venerable age. Only small specimens can usually be included in a collection, sadly in the wild there is only a small number of young plants as most were found, dug out and sold as decorative plants a long time ago. Barranca de Tolantongo where photograph No. 20 was made, is one of the places selected to protect these cacti for the future, the slopes of deep canyons being inaccessible to collectors. Fortunately, the plant is now available from nurserymen. In cultivation, the white hairs have to be protected from water, otherwise they turn grey and unattractive. Another reason for the relative rarity of this cactus in collections is the way the roots will rot if the plants are overwatered. Seeds are not always easy to obtain, as the huge columns are impos-

15

16

Distribution of the genera
Coryphanta and *Escobaria*

17

18

19

20

sible to climb and to fell the monumental 'cactus organs' as was done until recently — hoping they contain seeds — is a barbarous act. The seedlings are easy to grow, they require sunlight and warmth and a dry room in winter.

Cephalocereus hoppenstedtii (Web.) Sch. (21, 22). The regions south of the town of Tehuacán in Puebla are among the most interesting desert landscapes in the world. Some of the hills have been spared man's activity, and they are covered by thousands of cacti. Among many other cerei of this area,

C. hoppenstedtii is the most remarkable. Hundreds of cacti can be counted on one hectare. A 10 m tall cereus weighs about 400 to 600 kg. Nature has created here an extraordinary paradox for in one of the warmest and most arid areas of the world, many dozens of tons of water are stored in the plants, withstanding drought, scorching sun and wind. Only man could threaten this impressive spectacle.

This species can be successfully grown from seedlings in a greenhouse. It requires a loam-based potting soil, warmth and little watering. Throughout the years, the combination of strong, reddish central spines and tough, white bristly spines becomes prominent. It is so far an almost unknown cactus in collections. As with the previous species, the seeds are difficult to obtain. A synonymous name, under which the species appears in older literature, is *Haseltonia columna-trajani.*

21

Echinocactus polycephalus Engelm. et Bigel. (23). Nowadays, the genus *Echinocactus* includes only about five species, while at the end of the last century, most spherical cacti were placed under this name. The majority of them have now been classified in other genera and given other names. Consequently, one plant sometimes has several names. A case in point: *Echinocactus ingens* can have many other names (*E. palmeri, E. grandis, E. visnaga,* etc.) and there is argument as to whether they are different species or just local variations.

E. polycephalus comes from the states of California, Arizona and Nevada, growing in their driest and warmest deserts. *Polycephalus* means multiheaded, which corresponds to stems growing in clumps. Even small seedlings develop strong yellow to red spines. Flowers are unknown in cultivation. Seedlings grafted on tougher stock grow well using *Echinopsis* cv. or *Eriocereus jusbertii.* Grow under glass with a maximum of sunshine, warmth and fresh air.

Echinocactus texensis Hopff. (24) is better known as *Homalocephala texensis.* This name and the plant's classification in an independent genus are based on the structure and colour of the fruit, which is red, and without spines and scales after ripening. The flat, disc-like bodies of this cactus can be found in semi-arid regions over huge areas. Thousands of them occur

22

23

24

from southern Texas to central Tamaulipas in Mexico. They usually grow among acacia shrubs (*Prosopis* spp., *Acacia* spp., etc.) in alluvial sediments of crumbled secondary rocks on flat ground. They are rewarding plants in cultivation if watered sparingly and kept dry most of the time. Given ample sunshine and warmth, they grow in greenhouses for many years. The flowers are fringed and have a reddish throat.

Echinocactus horizonthalonius Lem. (25) is the smallest representative of the genus, adult specimens attaining 15 to 20 cm in diameter, and usually remaining disc-shaped. Unlike the previous species, it is difficult to grow. It requires a loam-based potting mix, maximum warmth and sunlight and minimal watering. This applies mainly to seedlings which are slow-growing but grow faster after grafting. Nurseries occasionally stock specimens which can be grown successfully. They flower on hot summer days, throughout the season. This cactus is distributed throughout the Chihuahua Desert, from San Luis Potosí up to the south of New Mexico and Texas.

Echinocactus grusonii Hild. (26) is a very popular cactus; thousands of its seedlings are sold every year. It is highly decorative thanks to the golden-yellow spines, fresh green body and regular spherical shape. It reaches about 60 cm in diameter, taking several decades to attain the adult stage and blossom for the first time. Some cultivated seedlings, however, produce flowers and seeds earlier.

E. grusonii is native to central Mexico. The photograph was taken on a mountain plateau, 2,000 m above the Moctezuma River, in the home of the columnar *Marginatocereus marginatus, Mammillaria elongata,*

25

M. herrerae. The *Echinocactus* are the globular plants in the foreground. In the past, *E. grusonii* was too widely collected and exported from Mexico as the decorative 'golden bowl'; as a result it is nowadays rare in the wild.

Echinocereus albatus Backb. (27). Most species of the genus grow in clumps, later forming mats of prostrate to creeping stems. This is one of the most tolerant North American cacti. *E. albatus* together with *E. reichenbachii, E. perbellus, E. purpureus, E. melanocentrus,* belong to the so-called pectinate (comb-like) species from the southern USA, mainly Texas, frequenting stony grasslands. They will withstand short periods of freezing temperatures, providing they are in a dry environment. Where winters are humid and low temperatures last longer, these cacti cannot be grown outdoors, although some experiments in which the plants were carefully shielded from damp in autumn and winter, were successful. The echinocerei are most outstanding in late spring when their large, colourful flowers open. The colour ranges from various shades of carmine to pale pink. The cacti later form large clumps. They can be encouraged to do this if wanted by beheading a plant which will then produce a number of young stems. An easy species to grow. The genus *Echinocereus* is also distributed on the peninsula of Baja California.

The denizens are among the most colourful and strong spined species, but their cultivation is not easy as they are very light demanding and also require warmth, careful watering and a loam-based soil. Interesting, robustly spined species are *E. ferreirianus* and *E. pacificus* with small pink flowers. The recently described *E. lindsayi* with up to 8 cm long stout spines is related.

Echinocereus pectinatus (Scheidw.) Engelm. (28 and 29) is a very attractive species for cultivation and is similar to other pectinate species; it flowers both easily and profusely. It is native to the Chihuahua Desert. The most beautiful variety is the var. *rigidissimus* (29).

Echinocereus fitchii Britt. et Rose (30) is a member of the group of pectinate echinocerei. In these cacti, the spines and areoles are arranged in regular, comb-like (pectinate) shapes. The bases of individual spines grow together and they can be detached from the areole in one piece. The spines usually form a dense cover of the ribs, in the dry season enveloping the body,

27

28

shielding it from the sunlight and preserving some humidity near the epidermis. New shoots have pink, brown and cream-coloured spines, later turning darker and making the base look ungainly. This is made up for by the formation of clumps: the old parts of stems shrivel and are covered by new shoots. *E. fitchii* comes from Texas and reaches to Oklahoma. Many other representatives of the genus *Echinocereus* are found in Mexico. A group of the so-called forest species frequents relatively moist forest sites, seeking shade under trees *(E. leonensis, E. morricalii, E. enneacanthus)* or open semi-deserts *(E. conglomeratus, E. ehrenbergii, E. polycanthus).* Of interest are the miniature geophytes, *E. pulchellus* and *E. knippelianus,* occurring in grassy meadows, sunk in the ground.

29

30

Distribution of the genus
Echinocereus

Echinofossulocactus dichroacanthus (Mart.) Britt. et Rose (31) is a spherical or short and columnar, relatively small plant with numerous narrow and undulating ribs. It usually flowers early in spring. It is native to grassy slopes and plains where, after the summer rains, grass covers the cactus stems, making them almost invisible. The species is easy in cultivation, needing humus-rich soil and regular watering in summer. Particularly robust spination can be developed in cacti placed in a greenhouse close to the glass for maximum light. The genus embraces some 25 species, but it includes dozens of published names most of which represent merely variable forms and local populations.

Echinofossulocactus coptonogonus (Lem.) Lawr. (32) is the most interesting, although a not very typical member of the genus. It forms few ribs and it is rather difficult to cultivate. It is a very robust cactus, in the adult stage reaching up to 30 cm in height. The flowers are inconspicuous as in other species of the genus. The petals have a purple-red to brown-green central stripe.

31

Escobaria runyonii Britt. et Rose (33). The genus *Escobaria* with its 12 species is characterized by its growth in clumps, dense, pale spination, and pitted seeds with red-brown testa (seed coat). Most species are only a few centimetres high and the flowers are inconspicuous, opening in spring and summer. *E. runyonii* and the related *E. bella, E. strobiliformis, E. muehlbaueriana,* are easy to cultivate and can be recommended for beginners. Other species, for example *E. orcutii,* are much more demanding and require an acid compost. The genus *Escobaria* is classified in the genus *Coryphantha* by some botanists.

Epithelantha micromeris (Engelm.) Web. (34) grows on limestone rocks in Texas and northern Mexico. The small stems growing in clumps are covered with dense, bristly, fine short spines. The small, pinkish or yellowish flowers arise from the crown. The fruits are bright red, cylindrical berries resembling those of the genus *Mammillaria.* The body tissues contain alkaloids producing hallucinatory effects for which they were picked and chewed by the local inhabitants.
Epithelantha pachyrhiza is a similar species with cylindrical stems and usually pinkish or yellowish spination. Both species can be cultivated in a humus-rich, limy compost. Grafting on low stock such as *Eriocereus* spp. is sometimes recommended.

32

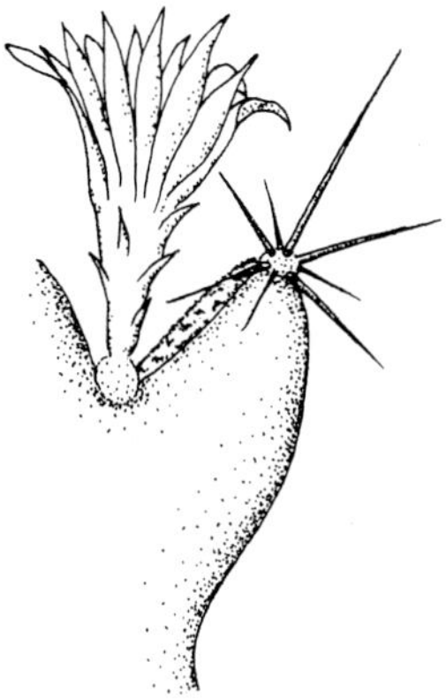

The flowers of *Escobaria* grow in the axil; a groove is noticeable on the tubercle.

33

34

35

Ferocactus histrix (DC) Linds. (35). The genus comprises some 35 species. In the USA, ferocactus is appropriately called Barrel Cactus, which describes the often considerable volume and shape of the plant. This species is not one of the largest, attaining only about 70 cm in height and width. It reaches the adult stage at about 20 cm and occurs in the states of central Mexico, growing on limestone. The seedlings are relatively tough in cultivation, but are slow to flower, though worth growing for their spines alone. Related and similar are *F. victoriensis,* flowering at 10 cm, and *F. glaucescens* with grey-blue skin and golden-yellow spines — one of the most beautiful species. Magnificent specimens can be seen in the famous Barranca de Meztitlán Valley, together with *Cephalocereus senilis* (19 and 20) and *Mammillaria geminispina.*

Ferocactus schwarzii Linds. (36) is botanically one of the more primitive of the ferocacti, closely related to the genus *Echinocactus.* It is similar in the structure of its ribs and areoles, and in the simple, pale yellow, open flowers and seeds. This rare species is found in a small area, in barely accessible mountains above Río Sinaloa. It is to be found on rocky cliffs, precipices and small plateaux on the steep slopes of mountain valleys. These places are covered with forest trees, and the presence of water is borne out by lichens and mosses, often growing on the older parts of cactus bodies. *F. schwarzii* is very rare in cultivation because its habitats are little known and difficult to get to. *F. alamosanus* grows further north and it has recently been introduced to cultivation.

Ferocactus latispinus (Haw.) Britt. et Rose (37, 38) is a small, flattened, exceptionally short columnar ferocactus with conspicuous, strong and colourful spines. The central spines are particularly well developed, being flattened and hooked. *F. latispinus* reaches flowering size at 10 cm, and since the flowers open late in autumn (buds appear in late September and October), they occur in cultivation only if the end of the year is sunny,

36

37

38

Distribution of the genus *Ferocactus*

then they will last till Christmas. They have an intense purple-red to brown colouration and open in succession. The variety *F. flavispinus* (38), with markedly strong, amber-yellow spines is also beautiful. In the wild this species can be seen abundantly in the area north of the capital of Mexico City in central Mexico.

Ferocactus steinesii (Hook.) Britt. et Rose (39) is one of the largest of the barrel-shaped cacti. Specimens over 2 m high, forming huge clumps and groups, can be seen in the western valleys of San Luis Potosí in Mexico. The reddish spination is prominent in the bushy vegetation, and the numerous 'heads' are seen from afar. The flowers are yellowish or pale orange. Many large ferocacti are found on the Baja California Peninsula; *F. diguetii* is one of the largest, exceeding 4 m in height. Seedlings can be successfully grown in slightly acid soil rich in humus. Only the Californian species cause problems and do not withstand temperatures below 10° C or a more permanent dampness of the compost. *F. chrysacanthus, F. coloratus, F. gracilis* and *F. johnstonianus,* belong to the most beautifully spined of all cacti.

39

Ancistrocactus scheeri (SD) Britt. et Rose (40) is a small spherical cactus with hooked central spines. In the wild it inhabits low hillsides of decomposed rocks in southern Texas and northern Mexico. Other related species such as *A. brevihamatus* and *A. tobuschii,* are also found in the southern USA. They are cultivated as for the genus *Astrophytum* and overwintered at temperatures of 5 to 10° C.

Leuchtenbergia principis Hook. (41). Plants growing under similar environmental conditions can develop remarkably similar forms, even though they are in no way related. This phenomenon is called convergence. In the case of *L. principis,* it is manifested by its similarity to the so-called rosette leaf-succulents, agave, aloe, echeveria and sansevieria. While these succulents have leaves, *L. principis* grows a rosette of tubercles (podaria) bearing areoles with spines, flowers and fruits on their tips. The plant is interestingly adapted to alternating periods of drought and humidity. In the wet summer, the rosette is open, and the sunlight can penetrate be-

40

41

42

tween the long tubercles; its energy is used in the tissues for photosynthesis. In the dry season, the cactus loses water, the tension in the tissues is reduced, and the podaria shrink into a compact, closed mound. The skin shrivels up and the plant cannot be overheated; further water loss by evaporation is also prevented. *L. principis* is native to the Chihuahua Desert, living in communities called *matorrals* with members of the genera *Agave, Prosopis, Fouquieria,* and other xerophytes.

Gymnocactus knuthianus (Böd.) Backb. (42) is widespread in central Mexico, in grasslands on low hills. It is a small cactus with dense white or yellowish spines, almost completely covering the stem surface. It flowers early, producing about 1 cm wide blossoms, during short days, that is in autumn and spring. It is easily cultivated and flowers freely. Other species of the same genus, for example *G. saueri, G. subterraneus* or *G. mandragora* present a more difficult problem for the grower, and some flourish reliably only when grafted. *G. horripilus* from Barranca de Meztitlán is an exception, as it does well on its own roots. Botanically, the genus *Gymnocactus* is closely allied to the genus *Thelocactus* (100—105) and to the genus *Turbinicarpus* (106 and 107).

43

44

Gymnocactus aguirreanus Glass et Foster (43). Two plants which do not correctly belong to the genus *Gymnocactus* are *G. aguirreanus* and *G. roseanus.* Their systematic classification is not clear, some members of the genus being reclassified as *Neolloydia.* Both species come from northern Mexico. They barely attain 5 to 6 cm in size. *G. aguirreanus* usually grows singly; it has bristly and prickly projecting spines. It does well in cultivation. The flowers appear in spring.

Hamatocactus setispinus (Engelm.) Britt. et Rose (44) is a small, spherical plant in cultivation becoming cylindrical with age. It usually grows solitarily with low ribs and hooked central spines. It is popular thanks to the innumerable flowers which appear throughout spring and summer on the crown. They are yellow with red-brown centres. The mature, bright red fruits are also decorative. *H. setispinus* comes from Texas and from the Mexican state of Tamaulipas.

Cultivation is easy and this cactus is a show piece for beginners. It can be grown in humus-rich soil, and during its period of growth can be watered freely.

Hamatocactus uncinatus (Gal.) Buxb. (45). Unlike the preceding species, this is a tough, drought-resistant cactus from the open, sun-scorched plains of the Chihuahua Desert. The bluish epidermis and long, strong and colourful spines make it an attractive quarry for collectors. On the other hand, the difficult conditions to which it has become adapted in its native home make it very demanding in cultivation, particularly in terms of maximum sunlight and warmth in summer, and of careful watering with long pauses.

45

Hertrichocereus beneckei (Ehrenb.) Backb. (46). The stems branch irregularly and do not form a crown of branches as do other cerei, creeping and trailing instead over rocks and into bushes. Young plants are covered by a beautiful, silvery-white coating of crystalline wax, which is in marked contrast to the jet-black spines. Only the tips of branches become less attractive, the older parts turning black, the downy coating being washed away by the rain, and the stems losing their handsome look. The flowers open in hot summer months. It is an attractive cereus for cultivation. If a hot and sunny environment is maintained and the stems are not sprayed with water, the white, waxy sheen will last for some time. According to our experience, flowers can occur when the plant reaches about 1 m in length. Seedlings take 4 to 5 years to grow to this size. This cereus is sensitive to excessive or permanent moisture in the soil, and does not tolerate temperatures dropping below 10° C.

46

47

Lophophora diffusa Croizat (47). More has probably been written about cacti of the genus *Lophophora* than about any other genus. This is because they contain psychically active alkaloids, mainly mescaline, which made the plants highly popular under the names peyote or peyotl. The Indian tribes used them in religious rites long before Columbus's arrival, and their popularity continued in the Christian era. The small, spherical cacti are still searched for and collected. Some botanists recognize two different species, *L. diffusa* and *L. williamsii.* They are very similar, especially in the juvenile stage and may be variations of one species. *L. diffusa* is, however, restricted in its occurrence to a small area of central Mexico. It inhabits the bottom of deep valleys, *arroyos,* eroded by drying-up streams, frequenting sandy sediments, but it also grows several kilometres farther onto the plains exposed to the scorching sun and wind, occurring in the shade of bushes. The flowers are usually white, rarely pinkish or yellowish. Because of the stout, turnip-like root, cultivation is less easy when the specimen is old. Porous potting soil, sparing watering and sufficient warmth are essential.

48

Lophophora williamsii (Lem. et SD) Coult. (48) is widespread over a large area of central and northern Mexico around the Chihuahua Desert and reaches northwards to Texas. It differs from *L. diffusa* in the composition of its psychoactive alkaloids. It is a variable species and has several scientific

49

50

names, it is however difficult to decide whether they denote different taxons or just very variable specimens.

Myrtillocactus geometrizans (Mart.) Cons. (49 and 51) translates as 'whortleberry cactus', which relates to the fruits which are numerous, small, red and look like whortleberries. In southern and central Mexico, they can be seen in markets, sold as edible *garambullos.* The cactus attains 3 to 4 m in height and forms a rich crown of branches covered by frosty looking wax. Up to 9 flowers can arise from a single areole, sometimes all opening together. Seedlings or cuttings of this species can be used as stock for grafting. Plants from its northern geographical limit are best because of their resistance to near freezing temperatures.

51

52

Lemaireocereus (Heliobravoa) chende (Goss.) Britt. et Rose (50) is a tree-like cactus with a trunk and crown of branches, and small nocturnal flowers. *Lemaireocereus (Polaskia) chichipe* and *Escontria chiotilla* are similar cactus trees.

Mexico is the home of many interesting cacti, highly valued by cactus growers, and the genus *Mammillaria* is undoubtedly the most important. It includes several hundred species, all small in size, colourfully spined and profusely flowering. The range of types is extraordinary; the genus has a vast geographical distribution as well as vertical distribution, from sea level to 3,000 m. The arrangement of the stem into tubercles increases the stem surface, and the dense spination with a marked seasonal change in body volume creates the shuttering effect already described. In favourable periods, the spines open up and the tubercles stretch out to absorb maximum sunshine, while in dry months, the tubercles are pressed together and the spines stick to the body. This and other progressive adaptations have made the mammillarias extremely hardy and vigorous cacti. Most of them are undemanding and do well in cultivation. The generic name — *Mammillaria* — refers to the most characteristic feature: the presence of conical, pyramidal or otherwise shaped tubercles (nipples). Another feature is the presence of areoles and axils with separate functions. The areole which bears the spines is situated at the tip of the tubercle, while the axil is either naked or covered with hairs, bristles or wool, and it produces flowers. The buds and fruits are therefore protected by spines from sunshine and pests. The fruits are usually juicy and red.

Mammillaria carnea Zucc. (52) was already being cultivated during the 19th century. It comes from the Mexican state of Oaxaca, occurring there in enormous quantities over a vast area; many forms can therefore be distinguished with a different number, colour and arrangement of spines. This cactus usually grows in large clumps and the flesh-red spination can camouflage the cacti in the thorny, bushy *chaparrals* where it grows. The flowers are red.

Mammillaria crucigera Mart. (53). The structure of tubercles, spines, flowers and fruits is very intricate, even the seeds are smallish. *M. crucigera* grows on the vertical walls of gypsum hills on the boundary of the states of Puebla and Oaxaca. Only tiny seeds can adhere to the clay particles on the steep walls. The size of seeds is therefore a successful adaptation to life in these particular conditions. In contrast, the species *M. napina* grows several dozens of kilometres farther away and its seeds are several times bigger. This plant is found on flat ground, where the large seeds can float in floods and be distributed throughout the environment.

53

Mammillaria boolii Linds. (54) occurs on the Sonora coast of the Gulf of California. It lives in humus and mineral-rich alluvial sediments. The related *M. insularis* from the adjacent islands differs in having stronger but less numerous spines and a stout turnip-like taproot. The third relative is *M. schumannii* from the southernmost tip of Baja California.

Mammillaria albicans Berg. (55). The southern end of the Baja California Peninsula in southern California is covered by a semi-desert vegetation. It is the home of some astonishingly beautiful and interesting xerophytes and succulents. It is also inhabited by numerous mammillarias. *M. albicans* has attractive pink flowers and white spination, only the tips of central spines being dark.

Mammillaria bombycina Quehl. (56) is a short columnar plant with decorative spines. It grows in clumps. Unlike the three previous species, it flourishes in more humus-rich soil and tolerates more frequent watering. It flowers in spring.

Mammillaria carmenae Cast. et N. de Cac. (57) was described as late as in the fifties of this century, and until recently was in fact unknown scientifically and inaccessible to collectors. The many attempts at rediscovering it in the wild ended in failure until the end of 1976, when it was found by Dr. A. B. Lau. Several months later, the authors of the book set out to search for it again in the wild mountains of central Tamaulipas. It is not easy to spot this beautiful species in its natural surroundings, even if you know

54

Distribution of the genus *Mammillaria*

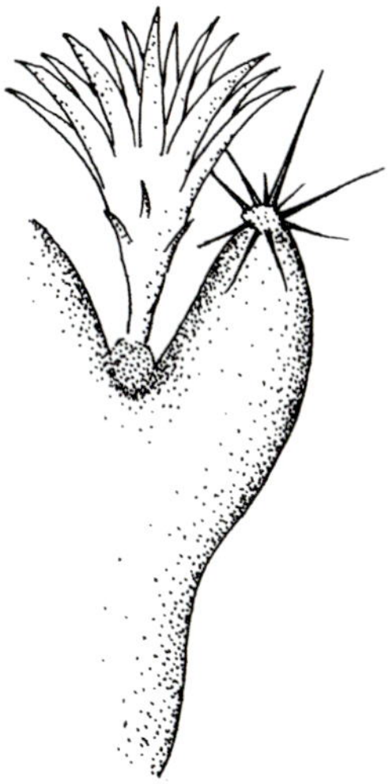

The flower of *Mammillaria* grows in the axil; the tubercle is without grooves.

55

56

57

the exact place to look, but the exhausting trip proved worth it. *M. carmenae* frequents relatively humid, partly sunlit cliffs at mountain summits. The golden-yellow tufts of fine, soft and supple spines virtually shine in the carpet of coniferous needles, moss and grass. The cactus flowers in late autumn, winter or in early spring, producing pale, whitish blooms.

Mammillaria candida Scheidw. (58). Few species are so popular and well known as this mammillaria, mainly because of its snow-white spines although there are variations in their shading, varying from yellowish to red and brownish. In the wild it is found on the limestone rocks of central Mexico. Some botanists classify it in the independent genus *Mammilloydia.* This cactus grows well from seeds which are easy to obtain. Loam-based potting soil with added grit is recommended for cultivation, together with minimal watering and wintering in a strictly dry environment at 10° C.

Mammillaria beneckei Ehrenb. (59). To see this beautiful plant, one has to travel to the watershed of the Río Balsas in the state Guerrero in central Mexico, where it is found in rather hot, broad valleys and connecting canyons, together with *M. guerreroensis.* Since the *cuenca* of Río Balsas is long, the plant is distributed over a large area, which results in a great variability of the stem shape, number of spines, and their dimensions and arrangement. That is why the plant has received many names over the years, notably *M. aylostera, M. balsasoides* and *M. nelsonii.* Some authors place it in the genus *Dolichothele,* others have created a special genus *Oehmea* because of its relatively large seeds with an interesting surface structure. The flowers appear at the height of summer: they are up to 3 cm across and of a simple structure and yellow coloured. The pear-

58

shaped fruits are bright red and adorn the plant longer than the blossoms. Cultivation is not easy; they need a constant temperature of 12—14° C, and the roots are sensitive to excess water. Growth from seeds is surprisingly fast, and within one season, the seedlings can attain up to 4 cm in diameter.

Mammillaria pectinifera Weber (60) comes from the flat limestone hills, the *mesetas* of central Mexico. It is better known in collections under its older name of *Solisia pectinata.* A related species, *M. solisioides,* grows some 100 km to the south. Both barely reach 4 cm in size and remain globular. Their spination is dense and white, and it is difficult to discover them in their home territory among the vegetation. A large part of the plant body is buried underground, and in the dry months they are totally retracted into the soil, leaving shallow pits on the surface; the plant can be discovered at the bottom. The flowers of neither species are conspicuous: in *M. pectinifera* they are white to pink, in *M. solisioides* yellowish. Their allies are

59

60

M. lasiacantha, M. denudata, M. magallanii and *M. roseocentra* inhabiting northern Mexico and extending to Texas. These plants remain small, with dense white spines, but they have striking red fruits appearing at the beginning of the growing season from the previous year's flowers. Their slow growth and difficult cultivation make both species rare as specimens. They do not tolerate a permanently moist compost nor decaying organic matter. For this reason they are often grafted, though this sadly causes changes in shape and size. Since they grow singly, they can be propagated only from seed.

Mammillaria dodsonii Bravo (61). Many species of this genus are large flowered, but the most remarkable are the relatively gigantic blossoms on miniature species of the series Longiflorae. It includes *M. dodsonii* and *M. deherdtiana* (67) from southern Mexico. Near the tops of 3,000 m mountains, above the timber-line, interesting plant formations, similar to the so-called *pára-*

61

mos, can be encountered. Scrubland vegetation is predominant here, and the rocky sites are bare or covered with grasses and succulents. In March and April, one can see conspicuous violet flowers reminiscent of crocuses, hidden in thick moss. Only on one's knees can one see that the flowers belong to 2—4 cm tall, densely spined cacti, buried in the ground and overgrown with moss and grass. The rainfall in these mountains is high, above 1,500 mm in a year, but the porous ground, strong, constant wind and scorching sun ensure that drainage is sharp and evaporation rapid which explains the occurrence of many other xerophytes and succulents. *M. dodsonii* responds well to cultivation providing a well drained acid soil is used with good ventilation. Under such conditions, it will tolerate ample watering.

Mammillaria dealbata Dietr. (62). In southern Mexico, one can often see small, white-spined, spherical, ovate or clustering cacti. They mostly belong to the Elegantes series, now named Supertextae. It has been found that the original *M. elegans*, which gave the series its name, cannot be identified with certainty. Some representatives of the series occur in dry and hot valleys at low altitudes, while other members frequent high elevations in the zone of pine or oak forests, seeking a moist environment under trees. The cacti mostly grow on limestone rocks or limy soil. *M. dealbata* specifically comes from the central part of the Mexican state of Oaxaca, from oak and pine forests. The trees hang with tillandsias and other epiphytes of the Bromeliad and Orchid families — all this proves the high tolerance of this mammillaria to humidity. There are no problems with it in cultivation, but it is necessary to water the white plants carefully to prevent the water from coming into contact with the spines. Any spraying makes the spines turn grey.

62

Mammillaria elegans DC (63). We shall use this name although, as mentioned above, there are doubts that it belongs to the depicted plant. The original description of 1828 is short and brief. According to contemporary criteria, it could cover several dozens of species. Perhaps it would be more appropriate to use the name *M. albilanata, M. lanata* or many others. But we believe that as the name *M. elegans* is still frequently used by growers it is better not to discard it until the botanists present their alternatives and a new name is accepted. We said that the complex of species around *M. elegans* could be successfully cultivated in temperate conditions, but there are several more difficult taxons, for example *M. dixanthocentron.* This species comes from crumbling slate rocks and marlstone. A loam-based compost with plenty of added grit should be used. In moisture retentive soil the cactus often loses roots. The same applies to *M. crucigera,* already described (53).

Mammillaria nejapensis Craig et Dawson (64). An attractive solitary cactus with deep green plant bodies, abundant wool like cotton wool in the areoles and axils, with hooked, downward-bent, long spines. It also comes from southern

63

64

Mexico, occurring in deciduous forests. In the dry winter season, that is from November to May, the trees are leafless and full sunlight can penetrate to the lowest layers of forests, reaching small cacti and other succulents. At that time, this mammillaria produces its woolly covering protecting the green plant body from sunshine. The first spring rains wash off the wool, and the plant passes the moist summer season, without the woolly coating. At this time, the trees and bushes again have their foliage, trapping a humid, hothouse climate on the forest floor below.

Mammillaria magnimamma Haw. (65). In many species, especially in the so-called green ones, where the colour of skin is visible through the overall effect of the spines, the fruits represent long-term decorations, often lasting longer than flowers. This is true of *M. magnimamma,* and of the related *M. compressa.* Both can be found in the vicinity of Mexico City. *M. magnimamma* is native even to the Mexico University campus, growing on young volcanic rocks. Cultivation is easy; a loam-based potting soil is suitable, moist in summer but dry in winter. Winter temperatures can decrease to 0° C.

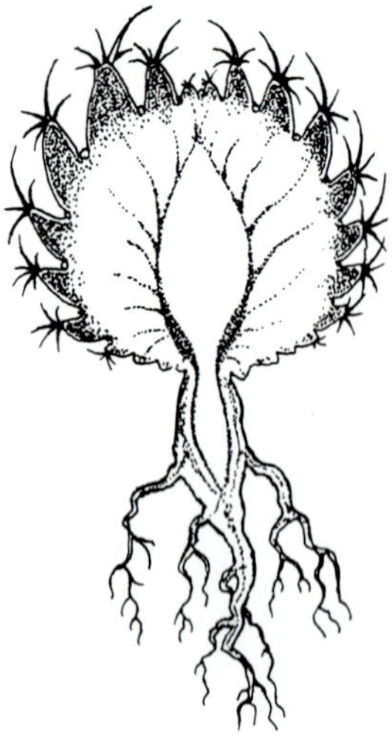

Cross-section of
a *Mammillaria* plant. Flower
axils are deep and protected
by tubercles and spines.

65

66

67

Mammillaria humboldtii Ehrenb. (66). In the village of Meztitlán, the inhabitants of lonely settlements in the mountains come to sell beautiful, densely white-spined specimens of *M. humboldtii.* It has over 80 spines to each areole. They will never tell where they find these valuable gems among cacti. No expert has therefore seen them in their natural environment. Many attempts have been made to find them high in the mountains around the village, but all have ended in failure. Judging from the requirements of the cultivated species, it can be deduced that the cactus comes from high altitudes and grows in black, humus-rich soil on limestone.

Mammillaria deherdtiana Farw. (67) may be called a twin of *M. dodsonii* (61). It is also found in similar altitudes. It has, however, different fruits and their ripening is different from other mammillarias. When the blossoms wither, the fruits remain hidden deep among the tubercles, in the axils. The seeds develop there and after maturing they stay in the body of the plant for many years. When the cactus perishes and becomes decomposed, the seeds are released and are scattered. An irregular and relatively low germination rate is common, but the seeds keep their viability for decades.

68

Mammillaria meiacantha Engelm. (68) represents the more northerly species. It grows in forest communities, frequenting deep sediments of gravel or sand, or in humus deposits. It grows singly in small colonies. Since the area of its distribution is large, several geographical races, forms and varieties are recognized. The picture shows a plant from the southern boundary of the area, from the flat hills in eastern Tamaulipas. Like other mammillarias, it has developed lactiferous ducts in its tissues, that is ducts containing latex, a milky, white liquid of unpleasant taste, repelling herbivores. When the plant is damaged, the latex flows out of the wound and coats it, protecting the cactus from infection and mould. *M. meiacantha* has many related species, some of them in the states of northern Mexico and south-western United States. The best known are *M. heyderi, M. mexicensis* and *M. glareosa.* One of them, *M. gaumeri,* appears also far in the south, in southern Mexico in Yucatan. *M. meiacantha* is a rewarding cactus for cultivation and can be recommended to beginners.

Mammillaria aureilanata Backb. (69). Two massive mountainous systems stretch across Mexico from the north southwards. The eastern range is called Sierra Madre Oriental, and its eastern side, exposed to the winds from the Atlantic, is covered with pine and deciduous forests. The mountain climate is cooler and more humid than along the coast or in the inland plateaux. The rainfall is more evenly distributed and occurs also in winter, even snow is not infrequent. This range is the homeland of many mountain mammillarias, e.g. *M. glassii, M. albicoma* and *M. plumosa.* In cultivation these grow through the winter period as well. The drier inland mountain range, facing away from the Atlantic, is the habitat of different species. They are also native to high altitudes (1,500 to 2,000 m), but they are less affected by rainfall, which makes them more resistant to drought, and winter is a period of rest for them, although many of them flower in these months, as do *M. dumetorum, M. schiedeana, M. subtilis,* etc. Their turnip-like roots will not withstand permanent moisture in the soil. *M. aureilanata,* from San Luis Potosí, is one of the most beautiful but also most delicate species in cultivation. It occurs in low limestone hillocks in crevices filled with black humus soil. A loam-based mix with grit and careful watering are recommended in cultivation. The winter temperatures can drop below 10° C, which delays the development of buds until the spring, when there is more sunshine and warmth and less risk of the buds or remains of flowers being attacked by rot as they can be in winter.

69

Mammillaria matudae Bravo (70). The Mexican states situated to the south-west of the capital are not desert-like in character, but many cacti can be found there. Botanists describe the local vegetation as a deciduous, dry forest. The closed, low-lying river valleys have a hot climate, dry for a long part of the year (winter months). As well as the tall, tree-like cerei, *Pachycereus* and *Pilosocereus,* and several members of *Coryphantha* and *Peniocereus,* the mammillarias can also be found here, mostly the species with long, cylindrical stems which form clumps. They inhabit steep terrain and the stems hang from the rocks or creep through the surrounding vegetation, reaching up to 1 m in length. They are related to *M. guerreroensis* from the Río Balsas basin, also to *M. matudae, M. meyrani, M. pitcayensis* and *M. magnifica.* They usually bear straight spines, but *M. magnifica* and *M. guerreroensis* can produce hooked spines in the juvenile stage or sometimes on aberrant stems. These plants require a warm winter (above 10° C). In the wild they grow in humus, and therefore in cultivation, one should choose a potting soil rich in humus and water them regularly, though they should be somewhat drier during cool spells, otherwise roots can begin to rot.

Mammillaria (Dolichothele) longimamma DC (71). Long, soft tubercles and sparse spination characterize this plant from moist, shady valleys of central Mexico. Among growers it is better known under the name *Dolichothele.* The experts, proving that the structure of seeds, flowers and tubercles differs from typical mammillarias, may be right, but if we consider that

70

71

there are many species forming a transitional stage between the two groups, the differences disappear. Cultivation is easy. Abundant watering and semi-shade are beneficial. The flowers appear throughout summer months.

Mammillaria herrerae Werd. (72) is one of the most beautiful, pure white-spined cacti. It grows on the bare limestone hills in central Mexico and it is sensitive to over-watering. It becomes mature when still very small and the open flowers entirely cover it. A white-flowered variety of this species is known.

72

73

Mammillaria parkinsonii Ehrenb. (73). Central Mexico is the home of two beautiful mammillarias with dense, strong, white spines. Later in life, both species form extensive, mat-like clusters of 'heads', often 50 cm or more in diameter. It is interesting that these formations are produced in *M. parkinsonii* by a division of the crown into two, then four, eight, etc. equally large 'heads'. This division is called dichotomous; it can be observed in many mammillarias, especially in the series Leucocephalae, which includes *M. parkinsonii.* In *M. geminispina,* the clumps are produced by side-shoots, tiny but developing branches which grow around the base of the stem and can be easily detached or cut off and used in propagation. *M. parkinsonii* has cream-white flowers, *M. geminispina* carmine-red flowers. Because both species come from steep limestone, marlstone and slate rocks, they are sensitive to over-wet compost. A porous potting soil should be used, the plant watered only when the soil is dry and on warm sunny days. Winter temperature can drop below 10° C.

Mammillaria solisioides Backb. (74). When travelling late in the year through the southern regions of Puebla and Morelos, a beautiful view of golden-yellow plains with sun-scorched grass and dark, contrasting volcanic mountains, steep hills and rocky walls spreads out all around. The stone-strewn hills hide the rare *M. solisioides,* which was named for its resemblance to *Solisia,* nowadays correctly called *Mammillaria pectinifera.* It resembles the latter with its dense, yellowish-white short spines, by its partially underground existence, and also by the structure and time of its flowers, and also the size, shape and structure of its seeds. The two, however, differ in several details, and their distribution does not overlap. *M. solisioides* has yellowish blossoms while *M. pectinifera's* flowers are pink. Both species pose problems in cultivation, mainly because of the difficulty of correct watering, long dry periods being necessary. Cultivation can be simplified by grafting, but grafted plants can take on some of the characteristics of the stock and instead of being small and spherical can grow large and elongated, ruining their attraction. This is often a danger with grafted plants.

Mammillaria patonii (Bravo) Werd. (75). Northwestern Mexico is the home of the so-called hook-spined, large-flowered mammillarias. *M. patonii* is widespread on the western coast. It is a forest cactus, found in crevices on rocky cliffs and walls above the sea. Fallen leaves form a coarse, acid humus, providing a habitat for clumps and thickets of cactus stems. In summer, they are packed with dozens of large, carmine-pink flowers. *M. sheldonii, M. iniaiae* and *M. occidentalis* are similar species.

74

75

Mammillaria pilispina Purp. (76). The western slopes of Sierra Madre Oriental, which face inland and are therefore drier, are the home of *M. pilispina.* We saw it on the gypsum hills, and were surprised to find the rock moist even in the drought. Gypsum easily weathers and turns into calcium sulphate which is highly hygroscopic and can absorb water from the humid air. For most of the year, the weathered rocks are viscous and sticky. The plants seem to enjoy these better conditions and apparently do not suffer from drought. On the other hand, the sulphate has chemical properties, not always beneficial to plant life and only certain well-adapted plants can survive here. The gypsum rocks are consequently always interesting and attractive hunting grounds for curious plants which differ from those of the surrounding area. *M. pilispina* does well in cultivation, even when grown in a normal soil free of calcium sulphate. On the contrary, it seems to do better than in the wild, its growth being more exuberant and it flowers profusely.

76

77

Mammillaria wilcoxii Toumey (77). The pine and oak-pine forests of northern Mexico and the south-western USA are the home of the interesting group of hook-spined, large-flowered mammillarias. In New Mexico, they ascend to over 2,000 m, to cool zones with temperature below 0° C in winter. Snow is frequent here but does not last long. The temperature drops usually only at night, while during the day the sun warms the air considerably. The frost has only short-term effects and the soil does not freeze. The mammillarias are found in the acid humus of decayed pine needles and oak leaves, and if they grow on a stony ground, it is always on acid granite. They have been regarded as difficult plants for a long time, and grafting has often failed. Some growers use acid peat and pine needles mixed with sand or crushed granite. The fleshy roots are very sensitive to over-watering and a porous mixture is essential, watering only when the compost has dried out.

Mammillaria sheldonii (Britt. et Rose) Böd. (78). The dark brown spines, with the central ones hooked, contrast with the broad petals of the flowers throughout its long summer season of bloom. *M. sheldonii* is very like *M. swinglei;* and may be one of its forms, differing slightly in the formation of the stem, number and colouration of spines in the areole, also in the arrangement and colour of the flowers. In cultivation, *M. sheldonii* requires more warmth in winter, more acid compost and sparing watering after being dried-off. Another related species is *M. mainae* from the regions of northern Sonora; it requires a porous, loam-based mix and maximum sunshine and warmth.

Other equally beautiful species are *M. microcarpa* from the south-western USA, *M. oliviae* from the borderline between Mexico and Arizona and *M. guirocobensis* from higher altitudes of Sonora.

78

79

Mammillaria theresae Cutak (79). The series Longiflorae is shown in photographs 61 and 67; it includes miniature cacti with very large flowers having long tubes. The northern distribution of this series is found in the states of Sonora and Chihuahua and includes *M. saboae, M. theresae* and *M. goldii.* All have to be looked for just below ground level where they remain during the dry winter period, retracting up to 1 cm underneath the ground. They were discovered in the sixties and seventies – quite recently and by chance. They are really visible only when flowering. Cultivation in a well drained potting mix is not difficult. Grafting is sometimes used to produce more blooms.

Mammillaria zeilmanniana Böd. (80). This brief survey of the rewarding and popular species of *Mammillaria* concludes with one of the least demanding and most easily cultivated of the inconspicuous dwarf cacti. The fresh green colour of the soft-fleshed body is wreathed throughout most of spring and summer with numerous small purple-pink flowers. This mammillaria can be grown on a window-sill and it will tolerate continual moisture in summer.

80

Normanbokea pseudopectinata (Backb.) Klad. et Buxb. (81) is a miniature cactus barely 2 cm long with a stout, turnip-like base. The fine white spines are not prickly, but densely cover the body and reflect the sun's rays. The flowers open in early spring before the plant begins to grow. This cactus lives on the flat grassy hills in San Luis Potosí. Both white and red-flowered populations are known, and are similar to the related *N. valdeziana.* Both species can be found in literature under the names *Pelecyphora, Turbinicarpus* and *Strombocactus.*

81

82

Neolloydia grandiflora (O.) Berg. (82) grows solitarily or in small clumps. The elongated stems are densely covered with glassy, awl-shaped spines. The flowers appear in high summer and they are carmine-pink in all species. Their distribution ranges from southern Texas, south to Hidalgo and Querétaro. A freely draining potting soil with added calcium should be provided for cultivation. The genus *Neolloydia* was founded by the American botanists Britton and Rose who classified within it the cacti which, like *Coryphantha,* have developed a horizontal furrow on the upper side of the tubercles. This formation produces blossoms, but unlike plants of the genus *Coryphantha,* these cacti bear small, dehiscent fruits and have seeds of a different structure. In these characters they resemble the genera *Echinomastus, Thelocactus* or *Turbinicarpus.* The position of the genus *Neolloydia* is still under review. The most handsome but difficult species is *N. matehualensis,* with a blue-green, waxy cuticle and prominent, strong spines. It is confined to a small area around the town of Matehuala.

83

84

Neolloydia conoidea (DC) Britt. et Rose (83) has dark central spines and white radial spines. It grows in many places in Mexico, and in each site the populations are slightly different. In Hidalgo and Querétaro, they have a tendency to grow elongated, cylindrical stems, a dark coloured epidermis and brownish spines forming large clusters. In the north of Nuevo León, the stems are more robust and the spination snow-white.

Opuntias form an enormous group of cacti with more than 300 species. Their area of distribution extends from Patagonia to Canada, from sea coasts to 4,000 m in the mountains. *O. ficus-indica* bears fruits which one can taste even in Mediterranean countries, and they are now exported to Europe. It is one of the exploited opuntias. Many others yield edible young shoots, *nopalitos*, which can be obtained only in Mexico. In the past, opuntias were introduced to many countries for their value as the food plant of the cochineal scale insect, *Dactylopius coccus*. In the 1880s, hundreds of tons of cochineal were made from them every year, 140,000 insects being needed to make 1 kg of the dye. They were also grown for hedging and some for fodder. Botanically, opuntias form a group characterized by stem segmentation. The segments can be flat *(Platyopuntia)* or circular in section *(Cylindropuntia)*.

Reduced leaves can clearly be seen in the areoles of young shoots. Each areole carries a large number of fine, hair-like barbs (glochids) which can be detached by the lightest touch and are very painful. The cells at the tip have arrow-like barbs making the glochid stick in the skin. Flowers of all species open during the day. They are variable in colour, mostly yellow, orange and red, and occasionally white.

Opuntia cantabrigiensis Lynch (84) comes from central Mexico. It is found on mountain plateaux in open country, exposed to full sunlight. It forms low, sometimes prostrate, clumps of blue-green pads, and early in spring can be covered with blossoms. In cultivation it remains neat and requires sunlight and careful watering.

Opuntia (Nopalea) dejecta SD (85) is a shrubby to low-growing, tree-like opuntia, developing a woody trunk with age and a crown of green, flat ellipsoid segments. With its related species (about 9), it is placed in the subgenus *Nopalea,* and distributed through southern Mexico and Central America along the western coasts. Since they are often used as decorative succulents in gardens and parks, they have been introduced to many tropical and subtropical countries where they have become naturalized. Their popularity is due to the production of striking red flowers of remarkable shape, lasting almost all year long. In European conditions, *O. dejecta* could be of importance for spacious interiors and as a summer park plant. It does not require any special care and tolerates the grower's mistakes.

Distribution of the genus *Opuntia* (including *Cylindropuntia, Tephrocactus,* etc.)

85

Opuntia (Cylindropuntia) tunicata Lehm. (86) is a decorative shrubby plant with dense, straw-yellow to pinkish spines enveloped in membranous sheaths — tunics — which can be easily removed with a nail. The segments are cylindrical and markedly elongated, especially the vertical ones. The lateral 'branches' are shorter and later form a crown. This opuntia ranges through a vast area covering Mexico, the Caribbean Islands, Colombia, Ecuador, Peru and Argentina. Many other cylindropuntias are known from the south-western USA such as *O. cholla* and *O. fulgida.* Outside their native lands, they are seen in larger collections, growing into decorative trees. Cultivation is simple, without special requirements, but they need plenty of head room.

Opuntia hystricina var. **ursina** (Web.) Backb. (87) is a highly decorative species with long segments and long, fine, silvery-white spines (hairs). It comes from south-western Arizona, from the arid areas of the Mojave Desert. Other varieties of *O. hystricina* are distributed in Nevada, Colorado and New Mexico, again in extremely dry deserts. In cultivation, this species is popular even in smaller collections, but the snow-white spi-

86

87

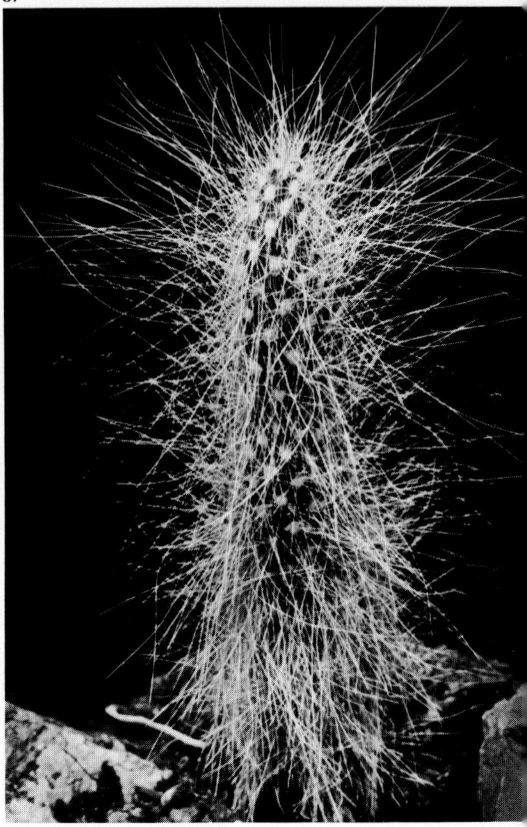

88

nation is more attractive on larger plants. In winter, it withstands temperatures below freezing point but is unlikely to prove hardy in damp, temperate winters.

Opuntia microdasys (Lehm.) Pfeiff. (88). Many succulent growers started their collections with this rewarding, indestructible plant. In the wild, this cactus grows into small shrubs composed of flat segments, but there are many forms differing in size and shape of segments and in the colouration of the glochid cushions. The glochids are yellow, red, brown, white. The cultivar 'Albatus', sometimes called 'Angel's Wings', can be particularly recommended to growers, its glochids unlike those of other forms being not prickly. As with most opuntias, *O. microdasys* can easily be propagated from detached segments, which root readily and rapidly grow new segments after planting. The flowers are of course a desirable ornament, but they usually appear on larger plants and are a rare phenomenon in amateur collections.

89

Opuntia pycnacantha Engelm. (89) is a strikingly handsome opuntia; its light spination is in contrast with red-brown areoles and the fresh green colour of the epidermis. It is a slow grower but less tolerant than the above-mentioned species. It requires careful watering and suffers if the compost is not well drained. It comes from the Baja California Peninsula. Of many other opuntia species, it is worth mentioning the species native to cooler regions of the United States and Canada. These are remarkably frost-resistant and can be grown in Europe on rocks, terraces and balconies. In June, they are a mass of yellow blossoms. The most famous ones are *O. phaeacantha, O. fragilis* and *O. compressa.* They serve as reliable stocks for the delicate scions of *Pediocactus* spp. and *Sclerocactus* spp. for growing outdoors.

Pachycereus weberii (Coult.) Backb. (90). The genus *Pachycereus* includes robust tree-like cacti weighing several hundred kilograms and attaining extraordinary heights (*P. gigas* and *P. weberii* can reach 11 m). Six species are distributed in western Mexico. They are more of interest for their immense size and age in the wild than as specimens for cultivation. We remember one day in southern Oaxaca; we were passing through a valley with the dead remains of hundreds of these giants. They are impossible to destroy with axes and machetes, so the local inhabitants wrapped them with heaps of dry twigs and maize straw, and virtually boiled the pulpy flesh. It was still many months before they could remove the cacti and cultivate the land for maize. Even if we suppress remorse and emotions, it is difficult to understand why people want to break the natural balance here, in the unfavourable conditions of deserts, where life depends so much on mutual harmony. The cacti, succulents or other xerophytes often cannot be directly exploited by man — they cannot be eaten or used as cattle fodder, or in industry — but they constitute vital components of the biological balance in arid regions. Their elimination threatens life, climate and reserves of underground water; erosion often follows and the land turns into a lifeless desert.

90

91

Obregonia denegri Frič (91) is a flattened, green cactus with its stem divided into numerous triangular tubercles. The spines are weak and inconspicuous, and the flowers arise from the top during the growing season. It occurs in only one place in the wild, in a broad valley near the settlement of Jaumave in Tamaulipas where it inhabits thick bushy forests, growing amongst fallen leaves in complete shade. In summer, heavy rainfall changes the desert climate into the sultry heat of humid tropics and the shield of leafy vegetation lets only sparse sunlight through, so that even on a sunny day, there is a permanent semi-shade. In cultivation, *Obregonia* requires a porous, loam-based mix and sufficient warmth in summer when it tolerates higher humidity. It needs a strictly dry environment at lower temperatures in winter. It is now usually classified under *Strombocactus.*

Cereus (Ritterocereus) griseus O. (92). Of the enormous number of columnar cerei, many are used in hedges; this grey-green species growing to 4 m in height and coming from the coast of Venezuela is one of them. It also occurs in the Caribbean Islands, mainly Cuba and Jamaica, and also in Mexico on the Atlantic coast. This cactus will even grow in places soaked with sea water as it tolerates a high concentration of salts in the soil.

92

93

Ortegocactus macdougallii Alexander (93). This cactus was discovered at the end of the fifties and described in 1961, a late date which is not surprising as it comes from the high mountains of the state of Oaxaca, where roads are scarce. The 5 cm tall plants with a white-blue epidermis and dark, protruding spines are found at the top of a single mountain with a different geological composition from that of the surrounding hills. Although the mountains around are covered with dense pine and oak forests, this hill is almost bare; the rock is covered only with sparse grasses, lichens and moss. *O. macdougallii* is abundant here: we counted 20—25 plants to 1 sq. m. It produces large, bright yellow flowers early in spring. It has not been specified yet where this cactus belongs taxonomically, and it still remains the only species in the genus. It is easily cultivated in a freely drained compost, providing it is wintered at above 12° C. Flowers are rare in cultivation.

Pilosocereus chrysacanthus (Web.) Byl. et Rowley (94). Together with the species which follows, this columnar cactus in adult age produces modified areoles at the top of its stem and branches which grow spines, hairs, wool and later flowers, and are known together as a cephalium. *P. chrysacanthus* is found in southern Mexico in relatively moist forest communities, as also is the following species.

Pilosocereus palmeri (Rose) Byl. et Rowley (95) comes from eastern Mexico. It has handsome seedlings with hairy areoles and light spines in juvenile stage, and they are very rewarding in cultivation. *P. palmeri* flowers readily when about 1 m high, even in a greenhouse. Wintering above 12° C is required. A normal potting soil is quite suitable.

94

95

96

Pediocactus (Pilocanthus) paradinei Benson (96). The genus *Pediocactus* comprises dwarf cacti from the American southwest, which live in grassy terrain. Their cultivation under glass is difficult, unless they are grafted on hard stock (*Trichocereus* or frost-resistant opuntias). They require plenty of warmth and sunshine in summer, and a cool to frosty, absolutely dry environment in winter. *P. paradinei* has pale, yellowish flowers. It is one of the most valuable and expensive cacti.

97

Pelecyphora asseliformis Ehrenb. (97) is a miniature cactus from the central part of the state of San Luis Potosí in Mexico. It has interesting areoles with spines no more than 2 mm long which remind one of tiny woodlice which have given the plant its specific name. Like the other representative of the genus, *P. strobiliformis,* it frequents stony sites and is sunk in the ground. The flowers of both species are carmine-pink and occur throughout the summer. Both cacti require much care in cultivation; plenty of warmth and careful sparse watering. They should be kept dry when temperatures drop and in winter.

Strombocactus disciformis (DC) Britt. et Rose (98) is another exceptional species with specific requirements both in the wild and in cultivation. Its natural habitat is the near vertical slaty walls of the canyons of the Río Moctezuma and its tributaries. Since it lives on shady northern slopes lacking sunshine, it has no protection from sunlight, lacking spination, any waxy coating on the epidermis, wool or hairs. It has the dust-fine seeds, necessary for the existence on near-vertical ground. Cultivation is difficult and it is recommended that a gritty and very porous compost be used. Watering must be applied carefully, and only during warm summer days, which is when it flowers. It is ecologically close to *Aztekium ritteri* (12).

98

Neobuxbaumia (Rooksbya) euphorbioides (Haw.) Buxb. (99) is a huge, columnar cereus with decorative areoles and dark spines. It is a slow grower suitable for larger greenhouses, living for many years. In winter, keep at temperatures above 10° C. The other 9 species of the genus *Neobuxbaumia* have fine, dense spines and they are rather sensitive to humidity and to temperatures below 10° C.

Thelocactus bueckii (Klein) Britt. et Rose (100). The whole genus includes some 12 species, mostly distributed in the Chihuahua Desert, in places with a hot, dry climate. They are usually solitary with the exception of *T. leucacanthus* (105), and have spherical, flattened bodies, tough tissues and prominent spination. Since they grow in mineral soil, their cultivation is not too easy requiring a well drained compost and careful watering only on warm summer days. They flower in high summer. They are easily propagated from seeds; the seedlings are fast-growing and if grafted will flower in 3—4 years. The generic name is derived from the Greek word *thele,* meaning tubercle, referring to the tuberculate structure of the ribs in all the known species. The colour of flowers within the genus ranges from several shades of pink to white and yellow, but as with some other cactus genera, it is not stable even within one species. *T. buecki* has pink blooms.

Thelocactus hexaedrophorus (Lem.) Britt. et Rose (101). The specific name *hexaedrophorus* is a Greek composite word referring to the hexagonal structure of the ribs or tubercles. The flowers are silvery-pink to white, or creamy yellow. It is found in San Luis Potosí in Mexico.

99

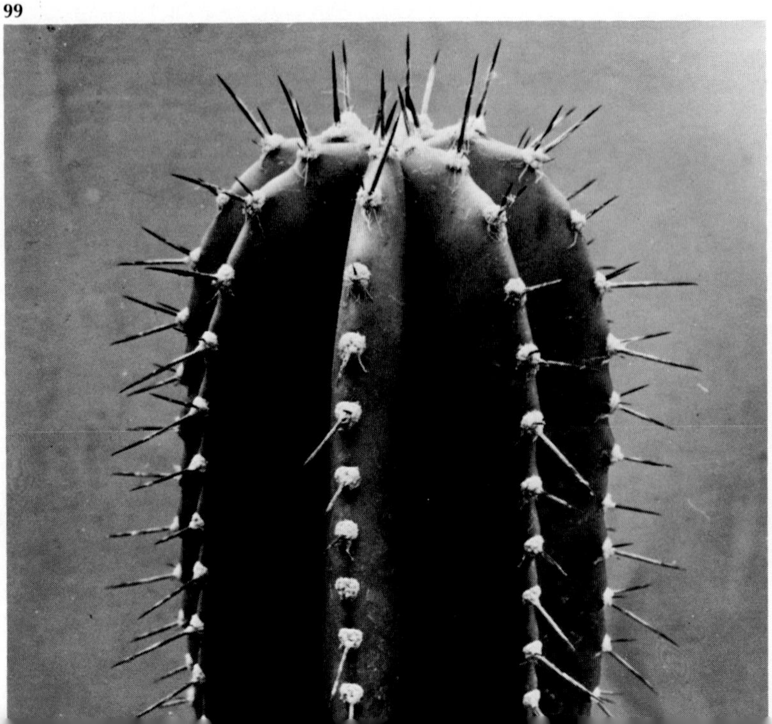

100

101

102

Thelocactus bicolor (Gal.) Britt. et Rose (102). Like in the preceding species, this is a small spherical plant with prominent spination. It comes from the flat, alluvial sediments of Texas and northern Mexico. Several local varieties are known; the most beautiful, white and dense spines are in var. *bolansis* from Sierra Bola. *T. schwartzii* from Tamaulipas is a related species, as is *T. heterochromus* with markedly strong hooked spines, from Zacatecas and Durango. *T. bicolor* and *T. schwartzii* are the easiest to cultivate under glass. In warm summer months, they tolerate ample watering, but must be kept in a dry place in winter.

Thelocactus matudae Sánchez-Mejorada et Lau (103) is the most recently discovered species of the genus. It was described in 1978 and named after Prof. E. Matuda, head of the Botanical Institute at the Mexican University in Mexico City, an authority in the study of Cactaceae. The new thelocactus comes from the state of Nuevo Léon. It frequents gypsum rocks, growing in very humid places covered with bushy forests rich in ferns and selaginellas. Cultivation of the new species is simple and it is to be recommended to collectors. With its brilliantly coloured, carmine-purple flowers, often larger than shown in the picture, it certainly deserves to become popular.

103

Thelocactus conothele var. **aurantiacus** Glass et Foster (104) is native to the valleys of the western Sierra Madre Oriental in central Mexico. Several other varieties are distributed through the region. Only recently, American authors evaluated the whole rich scale of forms and varieties of *T. conothele,* which differ widely particularly in the density of their spination. The flowers are purple and yellow. Anyone who has not had the opportunity of observing the plants in their natural environment, must find it difficult to believe that all are variations of one species. Another group of densely spined species with bluish to green epidermis occurs in northern Mexico. The best known representative is *T. nidulans* with thick, fringed spines. It prefers a well drained soil with a high calcium content. In the wild it grows in full sunlight. Such conditions are difficult to simulate in cultivation. When the plants are small, they are less attractive as specimens, but are well worth growing and will flower within about 4—5 years. The cultivation of seedlings requires patience because they are slow-growing at the beginning and some growers like to speed this up by grafting.

104

105

Thelocactus leucacanthus (Zucc.) Britt. et Rose (105) is an interesting cactus from the mountain plateaux of central Mexico. It forms large clumps of fresh green stems. This clump-forming habitat is valuable in grass communities where the dry grass accumulates and there is always a risk of fire. The margins of clumps may be damaged by the flames but the central heads remain untouched. *T. leucacanthus* has pale, yellow-white spines and yellow flowers. Plants from the western part of its distribution area have purple-red to pink flowers. They have been classified as *T. leucacanthus* var. *schmollii* and var. *sanchez-mejoradoi.* All these forms and varieties are easier to cultivate than the northern species.

Turbinicarpus laui Glass et Foster (106) is a dwarf, only 2—4 cm high, and spends part of each year underground. There are about 6 species in this genus and they are related to the genera *Thelocactus, Neolloydia* and *Gymnocactus.* All are adapted to this partially underground life as is *Pelecyphora.* Most of the body is made up of a turnip-like base which shrivels up in the dry season, reduces its volume, and the plant is retracted into the soil. After the rains, it absorbs water and gains in volume, pushing the green crown above ground level. As with the other species, the flowers are seasonal, appearing only in spring.

Turbinicarpus pseudomacrochele (Backb.) Buxb. et Backb. (107) lives at the southern edge of the distribution of the genus, and it has become very rare. The flowers are usually pink, although white-flowering specimens are also known. Its close relative, *T. krainzianus,* has yellow flowers, similar to those of *T. flaviflorus, T. pseudomacrochele* and *T. schmiedickeanus.* All the species in cultivation are sensitive to watering; they require a well drained compost and longer dry periods in summer; in winter, they tolerate temperatures below 10° C. They can be easily cultivated from seeds and within 2—3 years will reach 1 cm in height.

Wilcoxia poselgeri (Lem.) Britt. et Rose (108). Some 10 species of wilcoxias have been described, but the classification of this close relative of the genera *Peniocereus* and *Neovansia* or *Echinocereus* is still under discussion by botanists. The cacti are low-growing and shrubby, with thin stems, circular

Flower position in *Turbinicarpus* — the flower grows from the areole.

106

107

108

in section, mostly with underground taproots for water storage. They range from Texas to central Mexico, occurring in shrub communities with members of the genera *Acacia, Fouquieria, Prosopis,* etc., using their branches for support, because their own stems will not stand erect. They flower in summer, with a succession of purple, violet-pink or white blossoms. Grafting is recommended in cultivation (Fig. 10). Various species of cerei can be used as stock. Grafting eliminates problems with the relatively delicate, turnip-like taproots, and the flowers will be more abundant.

Wilcoxia albiflora Backb. (109) is one of the dwarf members of the genus, with stems barely 20 cm long. It resembles *W. schmollii* with hairy areoles. On the other hand, in *W. striata* or *W. viperina,* the branches measure up to 1.5 m.

Sclerocactus whipplei (Engelm. et Bigelow) Britt. et Rose (110). The American southwest is the home of many rare cacti. Thanks to the extremely dry, sunny weather and specific geological conditions, small populations of specially adapted species have developed there. Their cultivation is extremely demanding. The conditions necessary for reproduction, that is producing

109

Cactus fruits are mostly pulpy berries, often edible and of excellent taste (*Wilcoxia* species).

110

flowers and after pollination seeds capable of germinating, are very difficult to simulate. To import these plants from their natural environment would be an act of vandalism; they are also included in the Red Data Book of Threatened Species and they are anyway unable to adapt themselves even in the microclimate of the best greenhouses. Seedlings are also very difficult in cultivation, but they develop well when grafted on *Echinopsis* cv. or *Eriocereus jusbertii,* and flowers can then be expected within several years.

In Europe there has been success in cultivating a second generation from seed. *S. whipplei* and *S. polyancistrus* are the most prominently spined cacti. *S. glaucus* and *S. pygmeus* are the easiest in cultivation. An important condition when experimenting with their cultivation is plenty of fresh air at night, because the northern cacti require low night temperatures for successful growth. Warmth and sunshine during the day are another must. Winter temperatures can be low; in their native habitat, frost is frequent, and they can sometimes be snow covered.

111

Chapter 2 CACTI OF SOUTH AMERICAN STEPPES AND MOUNTAINS

The South American subcontinent is as rich in cacti as are Central and North America. The South American cacti can be divided roughly into two main ecological groups. The first group includes the inhabitants of high mountain systems of the Andes and Cordilleras. Their western slopes are arid, and the Atacama Desert on the borderline between Chile and Peru is one of the most totally dry zones in the world. These mountain cacti are therefore mostly adapted to withstand the unbroken rays of the sun, drought, heat and extremes of temperature. Their bodies are tough and the spines are often strongly developed. In many ways, they resemble the North American cacti from the southern deserts of the USA and the Chihuahua Desert in northern Mexico. On the other hand, the eastern, flat part of South America, to the south of the great tropical forests, is covered with dry, deciduous woodland *(catinga),* and open grasslands *(pampas* and *savannas).* The foothills of the Andes further north are marked by even more humid forest-type communities and the local cacti require more water than the Mexican ones or those from the west of the Andes. They not only withstand higher rainfall or watering in their growing period, but they are more resistant to fungi and bacteria and tolerate a more humus-rich soil. Systematically, the South American cacti are not of the same genera as those of the north. A few exceptions confirm the rule—for example the genera *Melocactus* and *Pilosocereus* are known in both Central and South America.

Arrojadoa rhodantha (Gürke) Britt. et Rose (111) is a shrubby cereus with branches only 3—4 cm across. All five species form clumps and are distributed in Brazil, on the plateaux south of the Amazon River. They occur in dry woods on stony, porous rocks. They contribute to the formation of the scrub layer, being partly shaded by the trees. In mature plants the branches at the end of the vegetative season terminate in a felted cephalium surrounded by a ring of longer bristles. This formation produces slender, fleshy flowers with a porcelain gloss. The following spring, the branches grow through the cephalium stretching out to form a new cephalium for next season's flowers.

112

Acanthocalycium violaceum (Werd.) Backb. (112) is the largest species of the genus, reaching a height and spread of 20 by 30 cm. The genus numbers some twelve species, distributed in the northern provinces of Argentina, in the eastern Andes, at altitudes from 500 to 1,200 m. They frequent slopes and valleys with sparse woods, scrub, and thicker undergrowth of grass. Water supply is relatively regular in the southern summer (November to March). The southern winter, however, is dry and rainless. *Acanthocalycium* is closely related to the genera *Echinopsis* and *Lobivia.* It has spherical, ribbed stems; the flowers are funnel-shaped and appear at the crown or in the lateral areoles. The flower tube and later the fruit are characterized by scales, spines and hairs. Cultivation is simple under glass. A nourishing and humus-rich potting soil is beneficial, watered freely in summer. Only the yellow-flowered species are more sensitive to excessive watering. Keep cool in winter (5—10° C). The flowering stage is in early spring and summer.

Blossfeldia liliputana Werd. (113) is the smallest South American cactus, fully grown and flowering before it is even 5 mm high. It comes from the mountain passes of southern Bolivia. It shows adaptations similar to those in the

113

114

Mexican *Strombocactus* and *Aztekium* which also grow on very steep slopes—the seeds of the genus *Blossfeldia* are very tiny and spherical, equipped with a large corky hilum. The spineless stems later form turnip-shaped, stout taproots and offsets to produce colonies of 50 or more miniature heads. During droughts, the cacti dry out to mere papery scales seemingly stuck to the rock. The flowers are about 50 mm across, and whitish; the fruits often mature even if flower development is incomplete (cleistogamy). Growing young plants from seed is tricky because of their size, and some growers prefer to graft young plants on to cerei. This assists growth and flowering and clumps of dozens of heads will develop shortly and flower profusely, though body shape is altered. In winter they tolerate temperatures around 10° C.

Aylostera (= Rebutia) pseudodeminuta (Backb.) Backb. (114) comes from the dry inland ranges of northern Argentina. It occurs at higher altitudes among tufts of grass and shrubs on steep slopes. In cultivation, it is a rewarding, easily flowered species. Like most members of the genus, it requires a cooler place in winter, because at temperatures higher than 14° C, it will continue to grow, using up the reserves stored for the development of buds and flowers in spring. Plants over-wintered in a warm place produce few flowers.

115

Aylostera (= Rebutia) albipilosa (Ritter) Backb. (115) is a small, globose, later usually clump-growing cactus with dense spination and distinct, pastel-coloured flowers. It is classified under the genus *Rebutia,* and the name *Aylostera* is mainly used as a subgenus. Some 20 species occur in northern Argentina and southern Bolivia, on stony mountain slopes among grass.

116

Aylostera (= Rebutia) spinosissima Backb. (116) is a thickly white-spined cactus from the Salta region in Argentina. The flowers appear early in spring. The older plants blossom more profusely than younger ones and they are often completely hidden under an avalanche of blooms. The fruits are minute and bristly, a characteristic trait of the cacti in the subgenus *Aylostera*. The most handsome species are *A. heliosa* with short, white, adpressed spines and orange flowers, and the similar *A. albopectinata* with red flowers. Both can be grafted. Cultivation of other species of this subgenus is usually easy if a loam-based potting mix, plenty of sunlight and fresh air in summer are provided. At this time of the year they can be placed outdoors. In winter keep cool at 10° C.

Buiningia brevicylindrica Buin. (117) is about 30 cm in height and 15 cm wide; it may grow offsets at the base and form clumps. It is one of the cerei and is related to the genus *Coleocephalocereus*, forming a lateral cephalium with hairy and woolly areoles producing flowers. The spination is yellow, up to 6 cm long and outstandingly attractive, especially on cultivated seedlings. There are three species in the Brazilian state of Minas Gerais; they live in the acid humus accumulated on terraces and pockets of granite rocks. They grow together with bromeliads, mosses and lichens, which indicates a heavy rainfall. Cultivation requires temperatures permanently above 15° C, and even in winter the plants should be kept growing slowly by occasional moistening and by keeping them in the light. Cultivated seedlings can be grafted on *Hylocereus* spp. and *Selenicereus* spp.

117

Dendrocereus nudiflorus (Engelm.) Britt. et Rose (118) is a tree-like cereus occurring only in Cuba. It grows up to 10 m and forms an extensive crown of bare, spineless branches and a stout woody trunk with needle-like spines up to 7 cm long. The white nocturnal flowers are abundant and are without basal scales as are the fruits which are apple-sized when mature. This cactus can be found on the northern Cuban coast where it lives in scrubby and deciduous woods, *manigua costera,* together with other cerei of the genera *Harrisia, Leptocereus* and *Pilosocereus,* it is however rare. It is of little importance for cultivation, but it can be used for outdoor decoration, although only in subtropical regions. It quickly reaches small tree size.

118

119

Browningia (Azureocereus) hertlingianus (Backb.) Buxb. (119) is a tree-like, candelabra-like cereus up to 8 m tall and 30 cm wide, with an azure-blue, wax frosted epidermis, yellow-brown spines and dark felted areoles. Its blue colouration makes it very decorative and a favourite in large greenhouses. In the wild, it occurs in southern Peru, in the valleys of the Apurimac and Mantaro Rivers. The genus *Browningia* is restricted to large Peruvian cacti. It used to be monotypic, having *B. candelaris* as its only member, then the authors included other, formerly independent genera, *Azureocereus, Castellanosia, Gymnocereus* and *Rauhocereus.* The original *Browningia candelaris* is a remarkable cactus: the stem is covered with long, strong spines, but on maturing, the plant produces only short-spined, candelabra-like branches. These fertile branches will not root and fail if grafted, this confirms the specialization of adult reproductive branches, regarded by some authors as an analogy of the cephalium in other cacti. Cultivation is easy when propagating the plants by seeds or cuttings. They tolerate a more moist compost and more regular watering during the summer season. Wintering is recommended at temperatures above 12° C, in an absolutely dry environment. To keep the blue colouration, spraying with water should be avoided.

Cereus peruvianus f. **monstrosus** DC (120). Although the name *peruvianus* implies that originally this cactus was discovered in Peru, the record has never been confirmed because it was never found again and the species is only known from cultivation. In addition to regular, evergreen branches, it has decorative, white nocturnal flowers like the related *C. hexagonus* and *C. jamacaru.* The first record of the form *monstrosus* dates from 1828. It has been widely distributed since then. This form produces flowers and seeds under normal conditions. Some of the offsprings then inherit the parents' appearance, and new monstrous deviations can be selected. They often differ in the length of spines, in their colouration and in the colour of the epidermis. Since they are decorative, easy and fast-growing cacti, they are in demand for greenhouses and interiors of homes, halls and patios. In summer, they can be grown in parks and gardens. *C. peruvianus* does not withstand frost and in cold climates has to be taken indoors for the winter.

The genus *Cleistocactus* with some 50 species is distributed over the eastern slopes of the Andes, from central Peru across Bolivia to northern Argentina. The shrubby, partly prostrate to creeping, mostly thickly spined stems form offsets at the base and grow in clumps. Some species mature early — often at 30—40 cm — (as *C. straussii*) and have rich, interesting, radially symmetrical, tubular flowers, in some cases pollinated by humming birds.

Cleistocactus wendlandiorum Backb. (121) has erect stems up to 2 m tall, with dense grey-white spines. It flowers readily in cultivation, producing orange-red blossoms, up to 5 cm long. It is one of the most widely cultivated cacti.

120

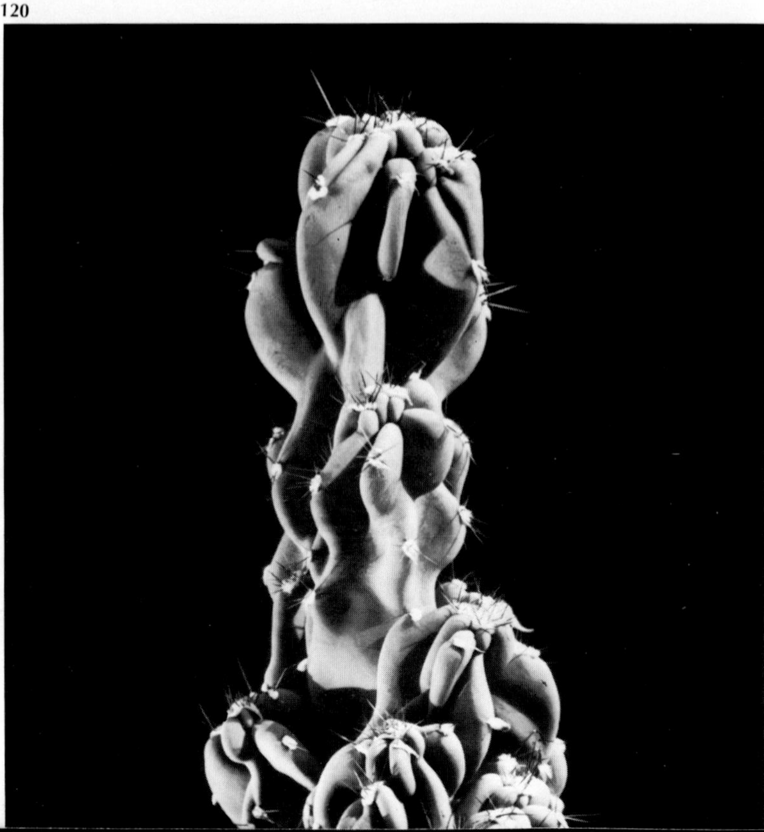

Cleistocactus candelilla Card. (122). In its homeland, around Cochabamba in Bolivia, it forms an undergrowth of at first erect, later prostrate stems, up to 2 m long. The stems have light to brownish, later greyish spines. The tubular flowers are red with white margins. *C. smaragdiflorus* has yellow to green blossoms. *C. straussii* and *C. jujuyensis* have the most beautiful, dense white spination.

Cleistocactus grossei (Wgt.) Backb. (123) is a coarsely spined species from the boundary zone of Argentina, Bolivia and Paraguay. It flowers profusely while still small, producing orange-red, relatively wide open flowers. *C. baumannii, C. colubrinus* and *C. flavispinus* are related species. All are cultivated in greenhouses where, if planted outside in pots, they rapidly mature into profusely flowering decorative plants. Later they can be shaped by pruning the older branches. They tolerate winter temperatures about 10° C. They are not too particular as regards soil.

The genus *Copiapoa* comprises some 45 described species: spherical, clump-forming to shortly columnar cacti. Their homelands are the arid zones of northern Chile, growing on barren stony slopes and the flat terraces of the western Andes. The flowers are yellow in all species, relatively small and simple-structured. The fruits are small and inconspicuous; the seeds are black. The genus *Pilocopiapoa* was established for a single species which differs in having hairy fruits instead of the hairless ones borne by the other species of *Copiapoa.*

123

124

125

Copiapoa cinerea (Phil.) Britt. et Rose (124) is the most typical and best-known representative of the genus coming from bare, sandy and rocky plains where it grows in full sun. Individuals have been found up to 1.30 m in height. The surface of the plant-body is covered with a chalky white, waxy coating, undoubtedly ecologically important; it reflects the sunshine and prevents water-loss through the epidermis. Among the South American cacti, *C. cinerea* is the hardiest species adapted to an extreme dry climate.

Copiapoa krainziana Ritt. (125). The globose body is 12 cm in diameter at the most, but it produces offsets and forms cushion-like clumps up to 1 m across. The epidermis is reddish to light brown, while the spination is grey-white to white, relatively long and protruding.

Copiapoa humilis (Phil.) Hutch. (126) does not resemble its massive relatives — the miniature bodies are hardly 2 × 2 cm in size and grow largely underground. The ribs are divided into bulges and the spines are fine. It flowers freely in cultivation (unlike the others, mentioned below), and flourishes without problem. Careful watering is recommended and plenty of fresh air. Other miniature species include *C. hypogaea* with stout, turnip-like roots, also *C. montana, C. longispina* and *C. tenuissima.* The dwarf spe-

126

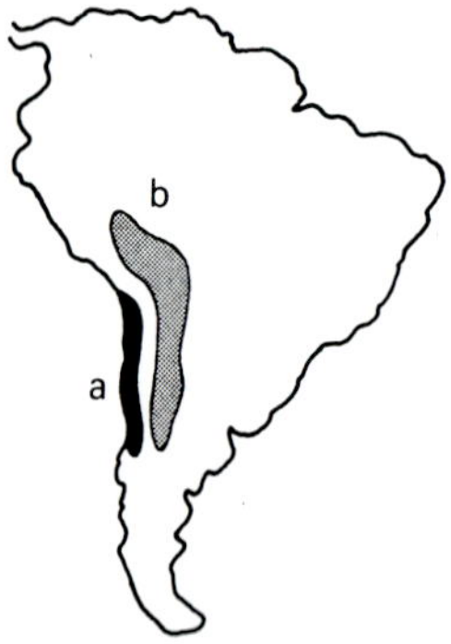

Distribution of the genera *Copiapoa* (a) and Lobivia *(b)*

127

cies have rather soft flesh and can be easily grafted, although they lose their original character even on a short stock; their internal structure becomes thinner, and the plants quickly get unnaturally big. Nevertheless, they flower readily and profusely. Some produce seeds after self-pollination — they are autogamous. Despite the fact that these plants come from open sites exposed to full sun, in cultivation, if situated near the glass in spring, they can be easily damaged by sun. Plenty of fresh air is therefore imperative.

Copiapoa columna-alba Ritt. (127) is probably only a variety of *C. cinerea* (124). It has a thicker stem, up to 20 cm across, a white waxy coating and black spines. The crown is enveloped in light orange felt, serving as a heat insulator. Readings taken in the wild showed that this felting prevents the young, embryonic tissues and developing buds from overheating.

Copiapoa dealbata Ritt. (128) grows in massive, semi-spherical clumps. It has elongated areoles. Most probably, it is another variety of *C. cinerea.* The wealth of forms of this species is far from exhausted. Other closely related species, such as *C. lebckei* and *C. gigantea,* lack the chalky white coating. The cultivation of these copiapoas is not always easy because of their adaptation to clear skies and brilliant light which can be obtained only in the optimal glasshouse conditions. The seedlings can be grafted onto *Echinopsis, Trichocereus* and *Eriocereus,* but the resulting plants do not resemble the natural ones and the chalky, waxy appearance is absent. Perseverance with seedlings on their own roots is the best way.

Only recently did southern and south-western Brazil reveal the richness of its cacti, and the genus *Discocactus* sprang from an unimportant one to a genus numbering some 30 described species. All the species are globose, flattened during drought. When mature, they terminate the growth of the stem and the crown produces at the tip a woolly formation, a cephalium, with scented nocturnal flowers. The terminal cephalium is characteristic of another genus — *Melocactus.* It is interesting that the area of southern Brazil with a high occurrence of *Melocactus* is also the home of the largest number of *Discocactus* species. Both genera have common qualities (body, shape, cephalium, surface structure of seeds), but there are differing traits as well, mainly the flowers. The theory that *Discocactus* and *Melocactus* are related is therefore acceptable; the flower ecology and morphology of each having developed separately.

128

Discocactus silicicola Buin. et Bred. (129). It is the easiest member of this genus to grow, quickly reaching flower-bearing dimensions at 7–8 cm. This takes about five years, provided sufficient warmth and nutrients are given. The very similar *D. magnimammus* is characterized by its pale hooked thorns and strongly developed tubercles. Both come from southern Brazil. Other handsome but more difficult species are *D. araneispinus, D. boomianus* and *D. albispinus* with snow-white spines. All were described in the seventies.

Discocactus albispinus Buin. et Bred. (130) also belongs to the complex which includes *D. araneispinus* and *D. boomianus.* The spines are white, the central ones up to 3 cm long and protruding. On reaching maturity, the cactus begins to form side-shoots, giving rise to numerous clumps. This group of species is characterized by relatively small, 3–5 cm long flowers, opening just for one night as do all discocacti. *D. albispinus* is found in the Serra do Francisco in the Brazilian state of Bahia. It is very similar to *D. zehntneri,* described from a specimen sent to Britton and Rose by post in 1921.

Discocactus horstii Buin. et Bred. (131, 132) is the smallest and most distinctive member of the genus *Discocactus.* It attains a maximum of 6 cm across and has short, adpressed spines. It comes from the Brazilian state of Minas Gerais where it lives buried in siliceous sands among grasses, moss and lichens. It protrudes for only several milimetres, during the humid period after the rains.

129

130

Discocactus is often mentioned in literature as a genus difficult to cultivate. We think the authors do not take into account the ecological conditions of the habitat—the cacti live in places with all year high temperatures and an annual rainfall of about 1,200 mm. This corresponds to the climate of an orchid house with minimal temperatures of 17° C, and watering throughout the year. The compost has to be porous, acid, loam or peat based. In an alkaline compost, these cacti lose roots. In summer, they tolerate permanent moisture providing the temperature is above 25° C.

Brasilicactus (= Notocactus) haselbergii (Haage) Backb. (133) is a globose cactus approximately 10 cm wide and high, thickly spined, with its crown slightly inclined to the sunny side. The flowers occur early in spring and sometimes a second time at the end of summer. The original forms had bright red

132

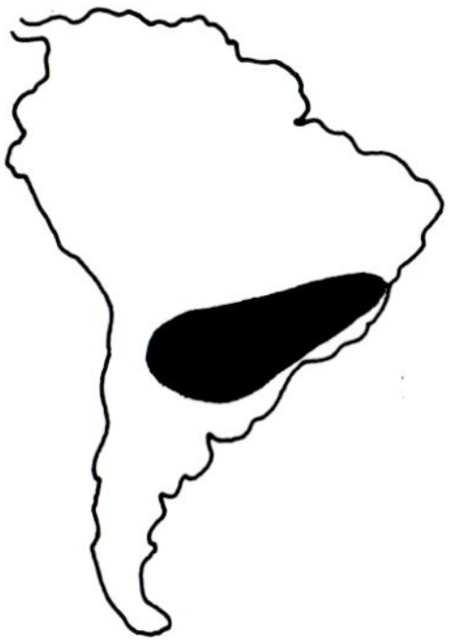

Distribution of the genus *Discocactus*

133

blossoms; yellow-flowered forms have since been introduced into cultivation. The cactus comes from southern Brazil; it is found on acidic rocks with moss and accumulated humus in cracks and on terraces. The closely related *B. graessneri* has golden-yellow spination and emerald-green flowers. The var. *albisetus* has white spines and yellow flowers and makes a linking form between the two species. Today authors place *Brasilicactus* in the genus *Notocactus* together with *Malacocarpus* (= *Wigginsia*) and *Eriocactus*. Cultivation is not difficult: an acid humus-rich potting mix is suitable with ample watering during the period of growth; cooler overwintering at 8—10° C is required. It is easily propagated by seeds, which are tiny and the growth is slow in the first year. Maturing is speeded by grafting on *Selenicereus*, *Hylocereus*, *Echinopsis* or *Eriocereus*, etc.

135

Denmoza erythrocephala Britt. et Rose (134). Both known species of the genus are massive, at first spherical (up to 30 cm in diameter), later columnar, up to 1.5 m tall, and many-ribbed. Spination is red-brown; in the adult stage, the areoles produce light, longish hairy spines. The red, tubular flowers are up to 7 cm long, and grow on the crown. *D. erythrocephala* and *D. rhodacantha* are distributed in western Argentina. They grow on slopes, on stony soils at altitudes of 1,000 m. Cultivation is not difficult providing they have a porous potting mix, plenty of fresh air and cool conditions in winter. The plant in the picture is about 50 years old, raised as a seedling. It is still adorned only by spines because flowers have not yet appeared though why this should be is uncertain.

Eriocereus pomanensis (Web.) Berg. (135) is a climbing, sometimes pendent cereus with 3–5 cm wide stems. Other species of the genus can have erect stems, grow prostrate on rocks or scramble through other vegetation. All have white, large and perfumed nocturnal flowers, resembling those of *Hylocereus* and *Selenicereus. Eriocereus pomanensis* is found in Argentina, Paraguay, Bolivia and Brazil.

136

Eriocactus (= Notocactus) leninghausii (Haage) Backb. (136) is spherical, later shortly-columnar and up to 1 m high. Often clump-forming, this densely spiny cactus comes from the boundary between Brazil and Paraguay. It is found in similar habitats to those of *Brasilicactus* and *Notocactus*, that is rocky and stony slopes in deciduous forests. These are rich in mosses and lichens which often grow on the bottom parts of the cactus bodies. In winter, the plants are exposed to drought and sunshine. The summer, however, is largely humid, with a rainfall of 1,000 mm a year, which means showers almost every day at a regular hour in the afternoon. Eight species are distinguished: *E. leninghausii* has elegant, slender stems and thick, golden-yellow spines, *E. schumannianus* (137) has thinner spination and a smaller number of ribs, but it grows in a columnar form to 1 m in height and 30 cm across. *E. magnificus* is the most recently discovered and most beautiful species, having a blue-green epidermis and golden-yellow spines. It matures when only 8 cm in diameter and like the two preceding species it has wide open, pale yellow flowers in high summer. The seeds of all species are tiny. *E. magnificus* and *E. leninghausii* form side-shoots at the base, giving rise to numerous clumps in the wild; in cultivation, these shoots can be used as a ready means of propagation.

Eriocactus (= Notocactus) schumannianus (Nic.) Backb. (137) is the most massive member of the genus. It forms no side-shoots and flowers rarely, which makes it a rare specimen in collections. Like other species of *Eriocactus* and *Brasilicactus,* it grows more on the side turned away from the sun, and the crown, on the contrary, becomes inclined to the sun. The phenomenon can be to a larger extent observed in plants suffering from lack of light or one-sided illumination. In the wild, the specimens growing in open country have almost straight crowns, while those of plants found in the shade under bushes are strongly inclined. Cultivation under glass is easy; it requires an acid compost, ample watering on warm days, and a dry place and temperatures of 12—14° C in winter.

137

Echinopsis aurea Britt. et Rose (138) is a spherical, later shortly columnar ribbed cactus, sometimes producing offsets and making clumps. It has yellow flowers which open before sunset. On cloudy days, the flowers will last until following day, but they wither during the morning on warm days. *E. aurea* comes from the Argentinian province of Córdoba, abounding on stony, grass and moss-covered slopes. The genus *Echinopsis* includes some 45 species distributed in northern Argentina, Bolivia, Brazil, Paraguay and Uruguay, mostly at lower altitudes, amongst grass and shrubs. Within the genus, the range of shapes and colours of flowers is very wide with the funnelform, elongated flowers of predominantly white colouration. Yellow or pink flowers are also known (139), and carmine-purple, orange and other red shades.

Echinopsis werdermannii Frič ex Fleischer (139) is a massive, bright green plant with strongly developed ribs, felted areoles and short awl-like spines. The colour of the flowers ranges from pale pink to pale red. This cactus comes from

138

the relatively humid and warm zones of Paraguay, from sandy soils overgrown with grass and shrubs. It resembles the best known species *E. eyrieesii* with snow-white flowers which can often be found on the window-sills of people who do not specialize in cacti. It prefers rather moist conditions, a rich soil, and shade from hot sunshine. By hybridizing various species of *Echinopsis* with other genera, mainly with *Lobivia, Trichocereus, Epiphyllum* and *Harrisia,* many attractive hybrids have been produced bearing colourful flowers. Hybridization and selection have given rise to cultivars of the genus *Echinopsis.* These are used as stock for grafting, mainly of seedlings. The cultivated species of *Echinopsis* will accept seedlings of all cacti, and no subsequent incompatibility of the stock and graft has been recorded. To grow cultivars it is important to know that they require a rich soil, moderate but continual humidity and semishade. The lowest winter temperature can drop to 0° C, but the Brazilian and Bolivian species need rather higher temperatures.

139

140

Eriosyce ceratistes (O.) Britt. et Rose (140) is a large, barrel-shaped cactus with a diameter and height of 1 m, comparable with the Mexican genera *Ferocactus* and *Echinocactus.* It is strongly and thickly spined, and the red to brownish flowers appear in wreaths on the crown. A single species is so far recognized, but due to great variability in the arrangement, length and colouration of the spines, many varieties have been described. In nature, this cactus is distributed in a wide area on the western coastal crests of the Chilean Andes. It can be found in valleys, at altitudes of about 200 m, or at elevations of over 2,000 m in the mountains. It is not easy in cultivation, seedlings, however, do well after grafting on *Eriocereus, Myrtillocactus,* etc., developing coloured, although shorter spination. Good light and summer warmth help with growth and a well drained compost is necessary.

141

The species of the genus *Espostoa* are shrubby to arborescent, thickly spined cerei, usually with white hairs. In adult age, they develop a lateral, woolly cephalium (like the Mexican genus *Cephalocereus).* They are found in Peru, reaching south to Ecuador. They can be encountered at 2,400 m, in dry mountain valleys of semidesert climate, like so many other cacti and in stony places in evergreen forests. Morphologically, shrub-like forms (for example *E. nana* or *E. melanostele)* and arborescent ones, with a candelabra-like top (*E. lanata* (142)) can be distinguished. These striking, snow-white cacti were first discovered by Humboldt and Bonpland in northern Peru and Ecuador at the beginning of the last century. In 1828, their finding was published as *Cereus lanatus.*

Espostoa nana Ritt. (141) is probably just a variety of the species *E. melanostele,* but it is smaller-sized, frail, with weaker stems.

Espostoa lanata (HBK) Britt. et Rose (142) is a tree-like type, attaining about 5 m when fully grown. The snow-white spination is mostly formed by the soft, hanging hairs, while the central, sharp, needle-like spines of glassy, white colour are not so easily noticed. Later, in adult specimens, they grow through the fine spines, reaching 3 cm in length. All members of the genus *Espostoa* are favourite decorative cacti. Even as seedlings, they can take their place in small collections. Larger specimens can be planted unpotted in glasshouses. They rarely mature into flower bearing dimensions under glass (picture 142 shows a flowering branch imported from the wild). *Espostoa* is not too particular as regards soil; it tolerates a humus-rich compost, moist in summer. Wintering is recommended at 12° C at least.

The genus *Haageocereus* (143) comprises some 50 thickly spined species with differently shaped stems. They come from both higher altitudes and coastal deserts, exclusively from Peru. They make colourfully spined specimens with strong vertical lines, indispensable in larger glasshouses. C. Backeberg divided them into Decumbentes, with prostrate and creeping stems; Versicolores, with interestingly coloured stems of shrubby appearance up to 3 m high; Acranthi, with robust

142

144

145

spines; Setosi, with bristly and hairy spines, and Asetosi, with combined spination. It is not a generally accepted division in view of variability of the species, but it is confirmed by other authors. The flowers are nocturnal: they open in the afternoon, bloom through the night, and close the following day at about 10 a. m. The fruits are striking red or yellow berries, naked or with traces of white wool or spines.

Haageocereus versicolor (Werd. et Backb.) Backb. (144). In the wild it is a shrubby cereus up to 1.5 m tall with varicoloured, yellowish, rusty and brown spination. Several varieties have been described, differing in the arrangement of spines, their length and colour. In cultivation, it is a decorative, rather resistant cactus, but it requires a dry place in winter and a warm and sunny situation throughout the year. Some species flower under glass.

Haageocereus acranthus var. **crassispinus** Rauh et Backb. (145) is a prostrate, thickly spined species from central Peru, ascending to altitudes of over 2,000 m. Spination is yellowish to dark brown.

Haageocereus chosicensis (Werd. et Backb.) Backb. (146) is one of the most frequently cultivated species, flowering at less than 1 m. Spination is fine and bristly in the juvenile stage; strong central spines grow later. The flowers are red.

Gymnocalycium is an interesting genus from the point of view of cultivation: it has approximately 70 described species. The cacti are spherical to shortly elongated, sometimes forming clumps, with naked, porcelain-like blossoms and conspicuous berry-like fruits. They range from the humid Brazilian plateaux *(G. horstii, G. fleischerianum)* across Paraguay to Bolivia, where many interesting species are found *(G. chiquitanum, G. marsoneri, G. hammerschmidtii)* to southern Argentina, the home of *G. gibbosum,* which grows in cool mountains.

146

147

Gymnocalycium denudatum (Lk. et O.) Pfeiff. (147) is a spherical, fresh green plant with spider-like, adpressed spines, white to pinkish flowers and large, black seeds. It comes from the humid Paraguayan savannas. It is a popular plant thanks to its easy cultivation. It requires a warm environment in winter, with a temperature above 12° C. Hybrids with interesting flowers, such as *G. netrelianum,* have orange and yellow blossoms, and *G. baldianum* produces pink to carmine flowers.

148

Gymnocalycium zegarrae Card. (148) is a representative of the hardier species; it comes from the Bolivian and Argentinian mountains. It has interesting, funnelform, sessile flowers. The large and globose *G. saglione,* reaching up to 30 cm in diameter, is a similar species from the Argentinian Salta, as well as *G. tilcarense,* for which C. Backeberg established an independent genus, *Brachycalycium.*

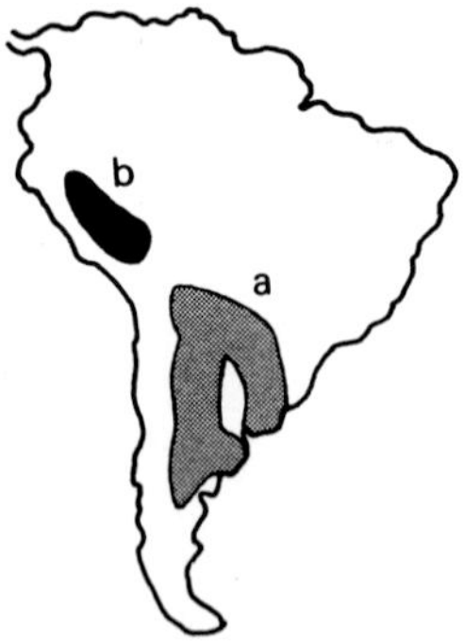

Distribution of the genera Gymnocalycium (a) and Matucana (b)

149

150

Gymnocalycium asterium Ito (149) is a flat, grey-brown plant with short spines and white, brownish-throated flowers. It is one of the cacti with helmet-shaped seeds, such as *G. quehlianum, G. riojense* and *G. vatteri,* from the Argentinian provinces of Salta, La Rioja and Córdoba.

Gymnocalycium spegazzinii Britt. et Rose (150) is up to 20 cm high and 15 cm wide. It comes from the stony slopes of northern Argentina. It has a highly variable spination. Seedlings are already decorative when small, but their cultivation is more demanding than in the other species, because they form stout, turnip-like taproots sensitive to over-watering. They flower when about 7 cm tall, later than the other species, producing 5 cm wide blossoms with pale margins and a reddish centre. The seeds are fine and black.

Gymnocalycium carminanthum Borth et Koop (151) is one of the newly found species, with striking yellow spination and red flowers. It is native to western Argentina, occurring in mountain valleys at 1,300 to 1,800 m. The red-flowered *G. baldianum* comes from more southerly mountains, and is tougher. All the mountain species are tolerant of a winter temperature about 0° C. In summer, they require plenty of fresh air; in dry and hot environments they aestivate (become dormant). They are not particular over the soil, doing well in a neutral, well-drained mix.

Gymnocalycium mihanowichii (Frič et Gürke) Britt. et Rose (152 and 153) is a spherical cactus, up to 12 cm high with sharp, transversely striped ribs and non-pricking

151

spines. It comes from the Gran Chaco region in northern Argentina, where it grows in sandy alluvial sediments in seasonal water courses. It is known in many varieties, several having white, pink or green flowers. In cultivation, it responds to temperatures below 10° C and to alkaline soil. It flowers freely at a size of 3—4 cm. From this species, several chlorophyll-less mutations have been cultivated.

Gymnocalycium mihanowichii 'Hibotan' or 'Ruby Ball' (152). In recent years, this has been undoubtedly the most popular cactus in the world. It is impressive thanks to its brilliant red colouration and perfect symmetry. It is a mutant, appearing first in 1941 in Japan. There are many mutations in which the plant loses the green colouring matter (chlorophyll), but they are deviations usually fated to die. They can exist only when grafted onto a stock providing nutrients. They mostly remain delicate even after grafting and rarely can be kept and propagated. Two things were important for the preservation of the chlorophyll-less mutation of *G. mihanowichii* — the vitality of the mutant and its formation of side shoots. Red mutants even flower profusely, more chlorophyll-free specimens grown from the seeds, later gave rise to other coloured deviations: yellow, orange, violet, pink, white and almost black. The colouration depends on carotenoids, pigments existing in normal plants (for example in carrots), but usually masked by the green of chlorophyll.

Gymnocalycium quehlianum (Haage) Berg. (154). This is the hardiest species of the genus, having an inconspicuous, flattened spherical, grey-green body, and producing cream-white coloured, pinkish-centred flowers from spring to winter. It remains flattened even in age, attaining the maximum dimensions of 7 × 7 cm. It comes from high altitudes in the Argentinian Córdoba province.

153

154

Islaya (= Neoporteria) grandiflorens Rauh et Backb. (155). This genus consists of approximately 10 species of spherical cacti up to 20 cm across, from the dry coastal areas of Chile and Peru. In certain systematic surveys, they are classified in the genus *Neoporteria* as a subgenus. *Islaya,* however, represents a largely homogeneous group, both in terms of appearance, spination, shape and colouration of flowers, fruits and seeds, but also in view of its existence in mineral, sterile, alluvial sediments, screes and sand dunes. According to ecological criteria, these cacti are extremely vigorous and adapted to extremes of drought. Thick and robust spination is a striking feature on plants growing in the wild; it serves as a protection from overheating in the open plains and slopes exposed to sun and wind. The flowers in the genus *Islaya* are yellow, but the petals are hairy on the outer side and have a central red stripe, or they are

155

entirely red from the outside, so that the closed buds seem to hide reddish blossoms. The flowers appear in summer depending on the weather, opening in succession. The fruits can appear even without pollination; they are then however a mere adornment, and without seeds. *I. grandiflorens* comes from the southern Peruvian border.

Islaya (= Neoporteria) minor Backb. (156) grows together with *I. grandiflorens* on the exposed windy slopes above the Pacific Ocean. For many days of the year, the only source of humidity is the thick sea fog *garua,* which condenses on the desert sand like dew. Cultivation of *Islaya* is not easy: it requires a well drained compost with a large percentage of sand, and a light place all the year round. Excellent results have been obtained with grafted seedlings which flower at 3 cm in diameter.

156

Eulychnia spinibarbis var. **taltanensis** Ritt. (157) is a tree-like cereus up to 4 m in height with relatively slender but strongly spiny stems. It occurs in the northern provinces of Chile, in the dry regions of the Chilean desert, occasionally ascending to higher altitudes. Some nine species are known, together with the many provisional names and numbered findings of F. Ritter. They include beautifully spined plants with long, varicoloured hairs. In cultivation, they can be used in larger greenhouses as ornamental, shrubby cacti which can be pruned by cutting into desired shapes. They flower when mature producing white blossoms.

Frailea castanea Backb. (158) is a miniature plant, barely 3—4 cm high with a red-brown to black epidermis and minute spines of black colouration. The flowers are yellow. This cactus is native to the border regions of Brazil and Paraguay, frequenting grassy, flat ground where it lives sunk in the soil. The genus *Frailea* comprises some 35, mostly miniature species with relatively large, helmet-shaped seeds, drifting in water and spreading over the flat, grassy savannas in summer floods. Pollination is interesting: it often takes place without the buds opening, that is by cleistogamy. The pollen grains mature before the full development of petals,

157

158

and since they are in direct contact with the stigma, they germinate and fertilize the eggs in the ovary when the flower is in the bud stage. In a certain number of flowers, the maturing of pollen grains is delayed, and the flowers open normally. This happens mostly in sunny, warm weather, and it is probable that in the genus *Frailea,* cleistogamy originated as an ecological protection from rain which would prevent pollination. The plants flower in summer which is characterized by daily rainfall.

In 1921, Britton and Rose published the name *Lobivia,* wittily inventing an anagram of the name Bolivia, to indicate the main centre of distribution of this genus. Many other species have been discovered in the neighbouring northern Argentina. Geographically, they all inhabit a well-defined area. The individual species often merge into each other (so-called cline species). Another interesting feature is that almost all the species occur in places with roughly identical ecological conditions — stony, grass and scrub covered eastern and central valleys and slopes of the Cordilleras, in places with irregular and low rainfall. It is difficult to give a brief description of the genus *Lobivia:* it includes

159

160

species with stems hardly 1 cm across, and prostrate, or huge, barrel-shaped cacti, 70 cm in diameter and 1 m high. Most of the species are small, single or clump-forming cacti with prominent spination. The flowers are red, yellow or white, mostly opening by day, even before sunrise, while the white-flowered species usually open their flowers in the evening.

Lobivia (Pseudolobivia) carmineoflora Hoffm. et Backb. (159) is a flattened plant with carmine-red flowers narrowing into a long tube. The flower shape is similar to those of *Echinopsis,* but the time of flowering is different: the flower opens in the morning and lasts one day and night. Because of this difference in timing one classifier has invented the genus *Pseudolobivia* including the species with features common to *Echinopsis* and *Lobivia.* Not everyone agrees, however, and if one believes that there are gradual transitions within genera then all can be included in one. This is why other authors now put the two genera together. However, both *Lobivia* and *Echinopsis* are rewarding cacti. They can be cultivated in less than ideal conditions, with reduced sunlight. They tolerate both drought and moisture, heat and cool weather. Favourable results are obtained in cultivation in the open in summer, but in winter, the plants have to be brought under glass at 5—10° C except in mild climates.

Lobivia (Pseudolobivia) kermesina Krainz (160) is a robust species, up to 15 × 15 cm, fresh green, with rusty brown spines and remarkable carmine flowers which open in the afternoon and persist through the next day until the evening. Flowers can be expected from May to September, when buds appear in succession. Cultivation is easy and generous feeding is recommended.

Lobivia jajoiana var. **nigrostoma** (Krzg. et Buing.) Backb. (161) is a small, bluish-green cactus with hooked central spines and a turnip-like taproot. The flowers in this variety are yellow, but var. *jajoiana* has pink and red blossoms.

161

Lobivia (Pseudolobivia) longispina (Britt. et Rose) Backb. (162) is a rather robust, strongly spined cactus up to 25 cm high, later producing offsets with white, diurnal flowers persisting till the next day. The spines are up to 8 cm long and the variety described by Kraiz as var. *ducis-pauli* has dark to straw-brown spines over 10 cm long. *L. ferox* is a similar species.

Lobivia (Pseudolobivia) ancistrophora (Speg.) Backb. (163) is a flattened, later spherical plant with spider-like, adpressed light spines. The white flowers open in the evening and last to midday of the next day. *L. obrepanda* is a related species with a massive, rich green body and white spines. The flowers are white to pink.

Lobivia winteriana Ritt. (164) is a small, brown-spined cactus having conspicuously large flowers with a long tube. The flowers appear continually from spring to autumn; they are carmine-pink. It is one of the recently introduced, attractive species, of easy cultivation.

162

163

Lobivia famatimensis (Speg.) Britt. et Rose (165) is a miniature cactus with many ribs and fine, light-coloured spines. The yellowish flowers with a short tube appear in June and July. It is a delicate species because of its stout turnip-like taproots, sensitive to permanent humidity. Like other delicate lobivias, it is a suitable subject for grafting onto *Cereus peruvianus, Trichocereus schinckendantzi, T. spachianus* and *T. pasacana.* The plant in the picture has been grafted. Many other lobivias have attractive spination, for example *L. kupperiana, L. staffenii* and *L. pentlandi,* or beautiful flowers, as *L. chrysantha, L. drijveriana, L. haageana* and *L. rebutiodes.* Hybridization with *Echinopsis* and *Trichocereus* has given rise to many interesting, colourfully flowered hybrids.

164

165

166

The mountains and deep canyons of central Peru are the home of small, spherical to shortly columnar cacti, with symmetrical (zygomorphic) flowers. The genus was named after the town of Matucana. Some systematists join this genus with others, for example with *Borzicactus* under which name it can often be found.

Matucana haynei (O.) Britt. et Rose (166) is a shortly columnar plant, up to 60 cm high with a dense mesh of whitish spines and prominent central, dark coloured ones. The crimson flowers appear on the crown. The fruits are naked. All the 20 or so species of the genus *Matucana* come from inland valleys and plateaux of the Andes, often ascending above 2,500 m. *M. haynei* was discovered at 3,200 m and *M. yanguanucensis* at 4,500 m! These situations are characterized by heavy rainfall, particularly in the southern summer (from December to April), and by a cool climate. The cacti can be found with clumps of grass and ferns, or in communities with shrubs.

167

Matucana aureiflora Ritt. (167). This species occurs in mountain meadows at an altitude of about 3,000 m. Adult specimens attain up to 30 cm in diameter and, unlike other species, they are flattened and the yellow flowers have a short tube. They resemble flowers of the genus *Oroya,* and come from northern Peru. *M. aureiflora* has radially symmetrical blossoms, unlike other matucanas. Cultivation of the mountain matucanas in the home is quite easy. They tolerate winter temperatures of about 10° C and should be given ample watering in summer and an acid potting mix with sufficient feeding. Plenty of fresh air and sunlight will encourage the development of beautiful, thick and varicoloured spination. The flowers are produced several times in spring, summer or early autumn. Some species, for example *M. aurantiaca,* with orange flowers, will blossom in short autumn days only.

Matucana yanguanucensis Rauh et Backb. (168) comes from extremely high elevations exceeding 4,000 m, where it grows in the rocks among moss and lichens. These places have good water reserves throughout most of the year. In cultivation, this matucana tolerates low temperatures and continual moisture. Only when sunlight is scarce does it tend to elongate too much (up to 1 m in height in the wild). *M. weberbaueri* with dense, bristly, yellow to rusty red spination and sulphur-yellow to pink flowers (var. *flammea*) is a very handsome species, as is also *M. comacephala* with a thick, snow-white mesh of spines and red flowers.

Matucana krahnii Donald (169). The plants from lower altitudes (up to 1,000 m) constitute an entirely different group. They grow in the deep canyons of the Maroñón River and its tributaries in Peru. They differ from their alpine allies in appearance, lacking the thick spination. They are rich green

168

and in cultivation demand a higher temperature in the winter resting period (temperatures below 10° C can be fatal). *M. krahnii* was described quite recently (in 1979). It comes from the vicinity of the village of Balsas, where when mature it forms clumps and groups on the stony slopes. *M. calliantha* and *M. formosa* are related species. *M. paucicostata* is a well-known species from this group, as well as *M. ritteri* which comes from higher elevations. All the above mentioned species have elegant, brilliant red flowers with long, narrow hairy tubes.

Melocactus is a very interesting genus; its first specimens were brought to Europe in the times of Christopher Columbus. The botanist J. T. Tabernaemontanus in 1588 called it a 'melon-thistle' — *Melocarduus.* Linnaeus called it *Cactus,* and this name was then applied to other plants and the whole family Cactaceae. On reaching maturity, at 5—10 years, the melocacti stop growing and form a cephalium in the middle of the crown, composed of closely set areoles in which the spines turn into bristles and a tuft of white wool. Each areole produces

169

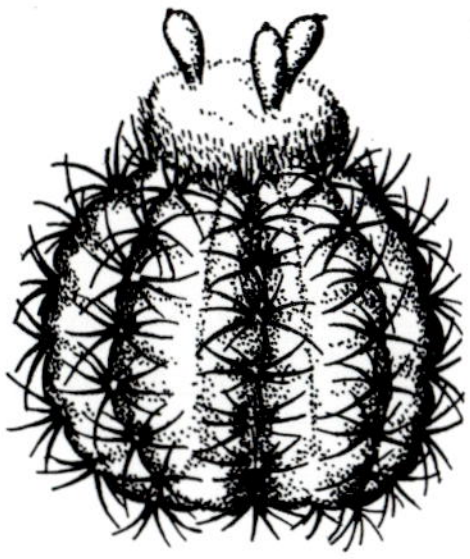

Cephalium in *Melocactus violaceus.* This formation produces flowers and fruits; other parts are immature and cannot flower.

170

a tiny flower. Within the genus, the colour of flowers varies only slightly, from carmine-pink to pale whitish-pink. In some species, the cephalium is hardly 2—5 cm high; in others, such as *M. communis* from Jamaica, it may reach 1 m in the course of many years. The distribution area of the genus covers almost the whole American tropical region.

Melocactus harlowii (Britt. et Rose) Vpl. (170) is very close to the larger species, *M. communis,* and is distributed in the Carribean Islands. It can always be found near the sea. In Cuba, it grows in limestone terraces covered with grass and low trees, in places exposed to full sunlight or in the shade under trees. The soil contains sea salts, which has led to the presence of specially adapted plants. In cultivation, *M. harlowii* is one of the rarest species, and very difficult, as is the following species.

Distribution of the genus *Melocactus*

171

Melocactus acunai Leon (171) also comes from eastern Cuba. In cultivation both species can be grown only as grafted seedlings — they are notable for their dense, handsome, yellow or orange, long spines.

Melocactus caesius Wendl. (172) is a spherical, bluish-green plant, reaching up to 25 cm in diameter, with reddish spines. The cephalium is composed of white wool and red bristles. This species is very abundant on the Venezuelan coast, from the Colombian border to the eastern frontier. It occurs in several forms and is also found on the nearby islands. Quite recently,

172

173

related plants were discovered in inland Venezuela, along the Orinoco River. They grow at low altitudes, together with several species of the genus *Pilosocereus* such as *P. claroviridis* and *P. moritzianus.*

Melocactus matanzanus Leon (173). Although it is also native to Cuba, it differs in many aspects from the other Cuban melocacti. Its distinguishing qualities are a greater tolerance as regards growing conditions, its habit of flowering at the size of 8—10 cm after only 4—5 years, and its ginger-red cephalium. According to our experience, it is a highly suitable species for cultivation, although the minimum winter temperatures must not fall below 15° C. Seedlings grafted on *Cereus peruvianus* and *Eriocereus jusbertii* flower in their fourth or fifth year.

Melocactus delessertianus Lem. (174) comes from the eastern coast of Mexico. Melocacti are usually described as difficult in cultivation, but if given warmth and light in winter, and humid heat in summer, they will flourish. However, they have to be divided into two groups in view of their tolerance. The most difficult to grow are the species from the Caribbean coastal regions (the *M. communis* group), and these reach the largest dimensions.

Equally delicate are the species from the eastern mountains of Peru and Ecuador, these include *M. peruvianus, M. equatorialis* and *M. bellavistensis.* The Brazilian species are, on the other hand, very rewarding such as *M. oreas, M. bahiensis, M. ernestii* and *M. grisoleoviridis,* although there are some exceptions. *M. azureus, M. diersianus* and *M. deinacanthus* are more difficult. In recent years, Horst and Buining have introduced many new Brazilian names. It remains to be seen how these plants will be evaluated after further study.

One of the South American genera which is rewarding in cultivation is *Neochilenia,* comprising small cacti from the northern Chilean provinces. Experts now usually include this genus within *Neoporteria,* although there are differences in the shapes of flowers, fruits, and in the dispersal of seeds. *Neochilenia* and *Neoporteria* are cultivated in a similar way, and they have the same requirements.

Neochilenia glabrescens Ritt. (175) is a miniature, brown-green cactus with a turnip-like taproot. In the wild it lives partly underground, in the same way as the dwarf *Copiapoa hypogea.* It flowers in summer, producing white to pink blooms, about 4 cm wide.

Neochilenia jussieui (Monv.) Backb. (176) has a dark green body with black or brown spines. The flowers are red. Since most species of the genus are very sensitive to too much moisture in their potting soil, it is sometimes recommended to graft them on low stock. This also speeds up flowering considerably.

174

175

176

Neochilenia pygmaea (Ritt.) Backb. (177) is a small cactus, 3—5 cm tall with striking, pink to cream-coloured flowers, native to coastal deserts. It thrives well when grafted. It tolerates winter temperatures below 10° C.

In many respects, *Neochilenia* is close to the genus *Neoporteria,* which is found in the same area in the south of central Chile, but it extends also to cooler regions. As they age, neoporterias become elongated, spination becomes protruding, prickly to finely hairy. The small-sized flowers are beautifully coloured and appear during short days, that is in spring, autumn and winter.

Neoporteria wagenknechtii Ritt. (178) is up to 30 cm high and 11 cm wide and is found in northern Chile. As with other species, it can be grafted on *Eriocereus jusbertii* and *E. martinii.*

Neoporteria villosa (Monv.) Berg. (179) is a columnar plant with a dark, brownish epidermis, white to black spines and hairs and purple-red flowers, with the petals white at the base.

177

178

179

Neoporteria nidus (Söhr.) Britt. et Rose (180) is similar to the preceding species. The pink or carmine flowers appear in winter. Other related and handsomely spined species are *N. gerocephala* and *N. cephalophora.* They may all be no more than forms of one variable species, coming from different habitats. All tolerate winter temperatures below 10° C.

180

181

182

Oreocereus neocelsianus Backb. (181) is a shrubby cereus 1—2 m high with thick white hairs enveloping the whole body, and yellow and brown-red central spines. It comes from northern Chile and southern Bolivia, ascending above 2,000 m. It grows in cracks and hollows on stony slopes and cliffs. It is a rewarding, attractive plant in cultivation, most suitable for spacious premises.

Oreocereus hendriksenianus Backb. (182) occurs in Bolivia and southern Peru. It forms large, dense colonies of slenderer stems than those of the previous species. In cultivation it thrives best after grafting on a stout stock. Two varieties, *densilanatus* and *spinosissimus,* are distinguished according to the density and length of spines. It is highly decorative and recommended for cultivation.

Oreocereus trollii (Kupp.) Backb. (183) has thicker and more colourful stems than *O. hendriksenianus,* and some forms with colourful central spines provide an excellent decoration. It is a slow grower both in nature and cultivation, even after grafting.

Of the South American cacti, the genus *Notocactus* is of great importance in cultivation. It includes spherical to shortly columnar cacti from the lower regions in the east, from Argentina, Paraguay, Uruguay and Brazil. *Notocactus* is a plant of grassland and forests with moderate rainfall in summer. All the cultivated species prefer an acid to neutral compost with an addition of humus, and moisture and warmth in summer.

Notocactus rauschii V. Vliet (184) is one of the most recently discovered species coming from the frontier between Brazil and Argentina, from stony hills at low altitudes. It belongs to a complex of species near to *N. mammulosus* and has very attractive spine formation. The similar *N. rutilans,* which has long been cultivated, has pink flowers.

Notocactus succineus Ritt. (185). Fine, white spination and relatively small yellow flowers place this species among the relatives of *N. scopa.* Cultivation is easy: the cactus does well in a porous, acid potting mix and benefits from liberal watering on warm days and wintering at temperatures above 12° C. Grafting of seedlings is sometimes recommended. Representatives of the subgenus *Malacocarpus (= Wigginsia)* are an interesting

183

184

group with sharp, tall ribs and a crown packed with wool. The recently described *N. buiningii* is the most beautiful species, needing a high winter temperature, while *N. erinaceus* or the frequently cultivated *N. ottonis* and *N. apricus* from the Uruguayan plains and hills are quite tolerant and will do well even in inexperienced hands. *Notocactus* responds badly to a high pH, that is to alkaline (limy) soil.

185

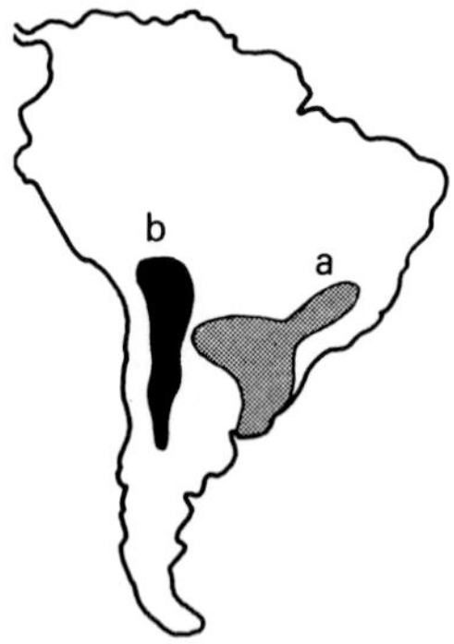

Distribution of the genera *Notocactus* (a) and *Rebutia* (b)

186

Notocactus uebelmannianus Buin. (186). A few notocacti have red flowers as has this species. The red colouration is probably the result of mutation in the make up of pigments, and the so-called red-flowered notocacti are otherwise identical to their yellow-flowered counterparts. The closely related *N. crassigibbus* and *N. arachnacanthus* come from southern Brazil as does *N. uebelmannianus,* frequenting warm, low situations in dry deciduous forests. Notocacti from Western Brazil, with hooked central spines, form another interesting group for which an independent genus *Brasiliparodia* has been suggested. *N. brevihamatus, N. bueneckeri* and their varieties are the best-known notocacti.

Only a few examples can be given of the wide range of South American opuntias *(Opuntia).* Growers should certainly pay more attention to the mountain genus *Tephrocactus,* which occurs in the Andes even above 4,500 m. Most tephrocacti can be grown in cultivation, but they do not attain their natural shapes and colours. The beautiful species with thick, white hairs such as *T. rauhii, T. malyanus,* and *T. floccosus,* are particularly difficult to keep in character.

Opuntia chakensis Speg. (187) is a tree-like cactus from the hot regions of the Gran Chaco on the boundary between Bolivia, Argentina and Paraguay. In cultivation, it quickly grows into a tree or shrub and flowers freely throughout summer. *O. elata* is a similar species. Both can be placed in the garden in summer months or in a border in the glasshouse permanently. Like most species with flat segments *(Platyopuntia),* they propagate easily by detached segments which root rapidly. The segments are sometimes used as grafting stock, but they contain a lot of mucus and handling is difficult. After the union is made, they become excellent stocks, especially for less vigorous *Cylindropuntia* and *Tephrocactus.*

The widespread and varied genus *Parodia* comes from the eastern mountains and foothills of the Cordilleras in Argentina, Bolivia and Paraguay. It consists of small, spherical to shortly columnar plants with dense, coloured spines. They flower freely at small size and grow well in cultivation. Botanists regard them as closely related to the genus *Notocactus,* but while notocacti cover the flatlands of eastern South America, parodias frequent mountainous areas where they have de-

187

veloped thicker spines and a reduction in seed-size. The Argentinian species in particular have very fine seeds, while the Bolivian species have larger ones. They are plants *par excellence* for growers; there are few more profusely and colourfully flowering South American cacti. They can be compared only to the Mexican *Echinocereus* and *Mammillaria.*

Parodia camblayana Ritt. (188) belongs to the group of species with long and coloured spines, more conspicuous than the smallish orange flowers. The group comes from south-eastern Bolivia: the best known is *P. maassii,* and other species are *P. camargensis, P. cintiensis* and *P. carrerana.* So many new species have been described recently, representing hardly more than differing forms of a natural population, that orientation in this avalanche of names is difficult even for an expert.

Parodia penicillata Fechs. et Steeg. (189) represents the small-seeded Argentinian species with beautiful flowers. It attains 70 cm in height and has thick white, yellow to rusty red spines. Cultivation is similar to the preceding species. It requires an acid compost, liberal watering in summer, and dry, cool overwintering. Propagation from seeds demands patience, but the growth of 5 mm seedlings is fast.

Parodia tilcarensis (Werd. et Backb.) Backb. (190) is a shortly columnar cactus with robust, hooked spines. It comes from northern Argentina and tolerates cool, dry winter conditions as do most cacti from this region, for example *P. stuemeri, P. jujuyana, P. uhligiana* and *P. rauschii.*

188

189

190

191

192

Parodia schuetziana Jajo (191) is a spherical species with fine, hooked spines and dark red flowers. It comes from northern Argentina. Of the other similar species, we can cite *P. sanagasta, P. mutabilis, P. aureicentra* and *P. catamarcensis.* These all need an acid compost but otherwise can be recommended for beginners. Propagation from tiny seeds requires patience.

Parodia echinus Ritt. (192) is a shortly columnar cactus with thick, straight spines. It is a native of Bolivia and, together with the related and similar *P. comosa, P. gracilis* and *P. mairanana,* can be recommended for cultivation. It flowers freely all summer.

Parodia subterranea Ritt. (193). The dark green epidermis and black spines in combination with the white wool of the areoles and red flowers make this species one of the most beautiful cacti. Cultivation is difficult because of the turnip-like taproot, and its sensitivity to continual humidity. *P. suprema* is a similar species.

The genus *Rebutia* comes from the same regions as *Parodia.* According to contemporary authors, the genus comprises *Aylostera, Mediolobivia* and *Cylindrorebutia.* The genus *Rebutia* embraces dwarf and small cacti, most living partially underground, in rocky crevices and moss cushions. Most species have thick, fine spination completely covering the epidermis. They bear remarkable flowers of various colours, appearing at the beginning of the vegetative season from older areoles.

193

Rebutia krainziana Kesselr. (194) is a 5 cm high plant, usually growing singly, with prominent white areoles and brilliant red flowers. Orange and yellow-flowered forms also exist. *R. minuscula* with an inconspicuous stem but frequent and abundant flowers is the best-known species.

Rebutia senilis Backb. (195) differs from the other species in its thick, white spines each up to 3 cm long. There are varieties with red, orange (var. *stuemeri*) and yellow (var. *kesselringiana*) flowers. Like most species of the genus *Rebutia* they are autogamous, that is the fruits and seeds are the results of self-pollination though in a collection hybridization is common. Cultivation of rebutias is easy: they need plenty of sunshine, water and fresh air in summer, and a porous compost. In the resting period, they require an absolutely dry environment and temperatures around 10° C. Only some of the recently imported species from lower situations in Bolivia prefer warmer winters.

Pterocactus tuberosus (Pfeiff.) Britt. et Rose (196) is a small plant with a large, water storing, turnip-like taproot, and thin, cylindrical, prostrate stems. Spination is fine, bristly and soft. The flowers grow terminally, at the ends of stems. The fruits have flattened seeds, differing from the seeds of all Cactaceae, and are hidden in the bodies. All seven or so related species come from southern and western Argentina, reaching as far south as the cool

194

195

regions on the shores of the Straits of Magellan. On their own roots, that is when not grafted, they are more difficult in cultivation because their turnip-like roots are sensitive to excessive moisture. A very well drained loam and sand mix is therefore recommended. Pterocacti tolerate temperatures around zero, but do not survive winter outdoors without protection. Systematically, *Pterocactus* is an opuntia, related to the genus *Tephrocactus.*

196

Reicheocactus pseudoreicheanus Backb. (197) is a spherical plant, 6–7 cm in diameter with red-green to brown-red epidermis, short, bristly spines and yellow flowers. In modern botanical systems, *Reicheocactus* is classified in the genus *Neoporteria.* It is also closely related to the genus *Lobivia,* particularly in the structure and location of flowers on the lower part of the body. The other two species, *R. floribundus* and *R. neoreichei,* have all the qualities of the genus *Neochilenia.* They come from northern Chile. The original habitat of *R. pseudoreicheanus* is unknown, and the cactus has never been found again. The plant in the picture has been grown for over 30 years in Europe. Grafting is recommended in all the three species, as with the miniature species of *Neochilenia* and cool, light winter conditions at 10° C should be provided.

197

198

Sulcorebutia arenacea (Card.) Ritt. (198) see text on page 182.

The genus *Trichocereus* embraces some 60 species of varied shapes, with elongated stems, robust spination and funnelform flowers. The largest species are *T. pasacana, T. santiaguensis, T. pachanoi;* they attain more than 5–8 m in height and form candelabra-like branches. There is also the small, 25 cm high, clump-forming *T. schickendantzii.* Trichocerei are distributed in mountainous areas of the Andes, from southern Ecuador across Peru, Bolivia, to Chile and Argentina. They are hardy in cultivation, fast-growing and highly decorative thanks to their spination. They need some time to reach their impressive dimensions.

Trichocereus fulvilanus Ritt. (199) reaches up to 1.5 m in height and forms large clumps of stems, some 7 cm across. The spines in some specimens are as much as 10 cm long. The flowers appear on the crown, they measure 10—12 cm and resemble those of *Echinopsis.* They are white and pleasantly perfumed. *T. fulvilanus* comes from the arid areas of northern Chile. *T. chilensis* is its close relative, forming thickets over a large area of coastal Chilean deserts. The largest species, *T. pasacana* from western Argentina, exceeds 10 m in height. It forms virtual cactus forests on the mountain slopes. The seedlings serve as excellent stock for all sorts of scions. Other species, for example *T. schickendantzii, T. candicans, T. chilensis, T. macrogonus* and *T. spachianus,* are all vigorous permanent stock cacti and tolerate temperatures below 10° C. They have one disadvantage: the graftings bear fewer and later flowers.

Trichocereus candicans (Gill.) Britt. et Rose (200) rarely reaches 1 m but forms large colonies of stems up to 12 cm across. It has a light epidermis and yellowish spines, up to 4 cm long. This cereus comes from western Argentinian provinces with stony hillsides. In collections it will over a period of many years grow into monumental, thick columns. Like other species, it

199

is not particular as to soil but benefits from plenty of fresh air and liberal watering. Smaller species flower freely and beautifully in cultivation, for example *T. crassicaulis* and *T. grandiflorus,* with bright red flowers.

The genus *Sulcorebutia* did not exist 20 years ago. *S. steinbachii,* known before 1960, was classified in *Rebutia.* After this date, many new species appeared, described under the new name. *Sulcorebutia* differs from *Rebutia* by its coarser spination, more elongated areoles and upward-facing grooves. Some taxonomists, however, refuse to believe in the independent existence of the genus *Sulcorebutia* and retain it in *Rebutia. Sulcorebutia* species form characteristic groups of small, often clump-forming, densely spined and interestingly flowering mountain plants. From the cultivator's viewpoint, they are beautiful flowering miniatures, flourishing especially after grafting when the risk of damaging the turnip-like taproots is eliminated. This is true of *S. rauschii, S. zavaletae* and *S. tarabucensis.* In winter, they tolerate temperatures falling below 10° C.

200

201

Sulcorebutia arenacea (Card.) Ritt. (198) is a handsome, spherical species with short, light coloured areoles and golden-yellow flowers. It is relatively delicate and grafting is often advised.

Sulcorebutia mizquensis Rausch (201) is a clump-producing cactus with tiny stems, light, adpressed, short spines and carmine flowers opening in May, as they do in most species. It has sensitive, turnip-like roots, and can be grafted on *Cereus peruvianus* or *Eriocereus jusbertii.*

Sulcorebutia lepida Ritt. (202). The small, dark green body rarely produces offsets; and the rigid spines are red-brown, up to 7 mm long. The flowers are red. *S. mentosa, S. tunariensis* and *S. totoriensis* are similar species. They can all be cultivated on their own roots in a neutral or acid compost, but to form larger clumps and more buds, grafting on reliable stock is recommended.

Vatricania guentheri (Kupp.) Backb. (203) is the only species of its genus. Some authors rank it in the genus *Espostoa,* which seems justified. *Vatricania* is a shrubby, densely spined cereus up to 2 m high. When mature, it grows a lateral cephalium composed of thick, rusty, fine bristles. It is native to the eastern valleys of the Bolivian Andes, to moderate elevations (800—1,000 m). Consequently, it does not tolerate cool places. As with other handsome cerei from eastern Bolivia and Peru, for example *Austrocephalocereus, Neobinghamia* and *Thrixanthocereus,* it is a decorative cactus for cultivation. It requires a permanent place with the temperature never dropping below 10° C, and sufficient moisture and sunlight in summer. The potting mix has to be neutral and porous. In such conditions, the seedlings will grow more than 25 cm a year and the cephalium can be expected when it reaches about 1 m. Mature branches with a cephalium will root more readily—side-shoots at the base are formed in the next season, quickly forming other branches with cephalia.

Uebelmannia pectinifera Buin. (204). This species was discovered only 13 years ago. There was quite a sensation when Horst and Baumhardt discovered this plant in the southern part of the state of Minas Gerais in Brazil. Five species are known today. Their grey-white surface reminds one of *Astrophytum* from Mexico and *Copiapoa* from Chile, although it is highly improbable that they are related. Recent studies have proved that their closest relatives are the South American genera *Notocactus* and *Frailea.*

202

203

Nevertheless, they remain very attractive but rarely obtainable cacti, even though thousands of them have been transported to world markets—and how many have survived? The imported plants make considerable demands on temperature: they have to be overwintered at above 15° C, they need permanent light and an acid to neutral compost. They flower rarely in cultivation, requiring the humid tropical warmth all summer, a position just beneath glass and a rich supply of mineral nutrients. Much better results have been obtained by grafting seedlings (204); *U. buiningii* bore flowers in the fourth year, *U. pseudopectinifera, U. meninensis* after 5 years, and *U. pectinifera* and *U. flavispina* after 6 years. The flowers are minute, hardly noticeable. They occur in winter among the crown spines, last briefly and measure little more than 12 mm in diameter.

204

205

Weingartia cumingii (Werd.) Marsh. (205) is a small, spherical plant with robust protruding spines and relatively small, naked flowers lacking hairs and spines. Some 17 species of this genus are known nowadays, coming from Bolivia and marginally *(W. vorwerkii)* from northern Argentina and Chile. They live among stones and on grassy slopes, together with members of *Rebutia, Lobivia, Gymnocalycium, Oreocereus,* and *Trichocereus.* Hutchinson placed them in the genus *Gymnocalycium,* where they remain in some recent classifications, but it seems they have not much in common, perhaps just the naked flowers. More conspicuous differences point to the Bolivian genus *Sulcorebutia,* where naked flowers, fruits, seeds and area of distribution are the same. Most species have profuse golden-yellow flowers, not too large but appearing throughout the summer in succession. Red-flowered species are also known—they were discovered in recent years on the Bolivian-Argentinian boundary. They are *W. torotorensis* and *W. rubriflora.* Cultivation of most species except *W. westii, W. fidaiana* and *W. vorwerkii,* also known as *Neowerdermannia,* is easy. A nourishing compost is recommended in cultivation, regular watering and a position in a sunny and aerated place.

206

Chapter 3. EPIPHYTIC AND OTHER FOREST CACTI

Cacti are found growing not only in arid deserts, semi-deserts or savannas and deciduous forests with scarce rainfall, but several hundreds of species have become adapted to life in humid regions. They can be encountered high in misty mountain forests, in gallery woods along rivers, or in evergreen, tropical rain-forests, at low altitudes. Because of the fierce competition of the other more massive and fast-growing vegetation, most cacti have settled in places where their succulence is advantageous. For instance, they inhabit tree tops seeking sunlight in places which dry quickly after rain, and where there is less competition. These cacti are called epiphytes — they live on other plants and do not put out roots in the soil, but attach them to tree bark, fixing themselves in crevices and the forks of branches. They do not harm their hosts and are not parasitic. Epiphytes do not live only in trees; they can be found on the steep rocky walls of canyons, or on solitary boulders sticking above the tree canopy. Among them are found the leaf-like cacti (Pereskoideae), which are hardly distinguishable from the leaves of other forest trees.

Aporocactus conzattii Britt. et Rose (206). In open oak groves in southern Mexico, above 2,000 m, one can encounter thick stems some 1.5 to 2.5 cm wide hanging from the forks between branches where they grow in accummulated humus. A sort of a suspended garden may be formed, with dozens of other plants and many insects or small vertebrates. Spherical cacti *Mammillaria elegans* or *Ferocactus macrodiscus* can also be found here, flowering several metres above the ground. Their seeds, like those of the epiphytic cacti, have been brought up by birds feeding on cactus fruits and seeds.

Aporocactus flagelliformis (L.) Lem. (207) comes from higher altitudes of central Mexico, where temperatures rarely exceed 24° C. It is the most common mem-

207

ber of the genus. It does well in the home, by the window, or outdoors in summer. The pot should be suspended to allow the thin, cord-like stems to hang down. Unlike other epiphytes, it tolerates alkalinity of the compost, but it does better in an acid, humus-rich soil. When overwintered in a warm room, occasional watering is recommended during winter.

Hylocereus undatus (Haw.) Britt. et Rose (208, 210). This is one of the best-known cacti, often planted in parks and gardens in subtropical areas. A small cutting situated at the foot of a tree or wall will soon become attached and climb up to reach the sunlight (208). In a favourable position, it begins to form many side-shoots. On a felled palm *(Roystonea regia),* these branches weighed almost 300 kg. The genus *Selenicereus* lives in a similar way in tree tops. The seeds of these plants often germinate in the soil, and the plants then climb up, attaching themselves by their roots, developing their crowns in tree tops and bearing flowers. The connection with the ground remains important — the plant must draw water and nutrients. It is therefore not a true epiphyte. In optimal conditions,

208

Primitive leaf-cactus *Pereskia aculeata* with flowers arranged in terminal inflorescences.

209

in tropical moisture and warmth, the growth of these semi-epiphytes is very fast; 3 m long one-year shoots are quite common. *Hylocereus* and *Selenicereus* are for this reason used as stock material. They are cut to 15 cm long cuttings, which quickly put out roots. After 2—3 years, the scions become too heavy and the stock has to be shortened or re-rooting carried out.

Selenicereus urbanianus (Gürke et Wtg.) Britt. et Rose (209) is a common epiphyte of solitary trees scattered over savannas and dry forests on the coasts of central and western Cuba and Haiti. It often inhabits rocks, covering them like a green carpet.

Selenicereus grandiflorus (L.) Britt. et Rose (211). This cactus, opening its huge white flowers at sunset, was named after the Greek moongoddess Selene, famous for her beauty. The authors remember a night in the Oriente Province in eastern Cuba, late in May. Large golden buds were hanging on solitary trees scattered in grassy pastures. There were dozens of them on each tree top. As dusk was falling, we encountered more and more open flowers — and in the dark, we saw shining white flowers like stars in trees. A sober estimate seemed impossible, but many thousands were blooming that night! The following day, only wilted remains hung from the trees and the fabulous scented flowers were gone. What force made all the buds blossom in a single night, will remain a mystery. *S. grandiflorus* is appropriately called 'Queen of the Night'. Some 25 species have been described, ranging from southern Texas throughout Mexico and the Carribean Islands, to the northern states of South America. The occurrence of *S. macdonaldiae* as far south as Paraguay is questionable. *Selenicereus* and *Hylocereus* can be placed in large, warm greenhouses, where they flower all the summer and can cover the walls. Other species include *S. hamatus,* with beak-like protuberances under the areoles, or *Selenicereus (Deamia) testudo,* with thick, 4-winged stems. Both come from Mexico and have equally beautiful flowers as those of *S. grandiflorus.* All need winter temperatures above 12° C.

Epiphyllum hookeri (Lk. et O.) Haw. (212) is a true epiphyte. The seeds have to germinate in suitable places in tree tops or on rocks. It is unable to climb as the preceding species do. In time, it forms extensive shrubs with stems

210

211

about 5—9 cm across which are coloured bright green. It comes from Central America and the northern part of South America, ranging northwards to Mexico.

212

213

Epiphyllum anguliger (Lem.) G. Don. (213) differs from the preceding species in having deeper notches on the branches. The flowers are smaller, 8–9 cm in diameter. Like all the other 20 species, they are open at night, possibly for the next day too. *Epiphyllum (Marniera) chrysocardium* is an interesting species with deeply notched, winged stems and spines on the ovary and flower tube. The epiphytic genera *Nopalxochia* and *Heliocereus* are similar, with white and red diurnal flowers. Together with *Epiphyllum* and *Selenicereus,* they are parents of many cultivated hybrids, better known as *Phyllocactus* or Orchid cactus and flowering in countless shades and colour combinations. While their parents need an acid, humus-rich compost and warmer overwintering, hybrid cultivars require a loamy, potting soil, summer temperatures not exceeding 28° C and semi-shade.

Wittia panamensis Britt. et Rose (214) has a climbing, twining, thin stem, reaching several metres in length and later hanging from tops and trunks of trees. It requires a warm and humid environment, and like most epiphytic cacti, is a decorative plant for epiphytic trunks, window-sills, etc.

Schlumbergera orssichiana Barthlott et McMillan (215) is a new species, described only in 1978. It is a handsome cactus with tipped segments and light carmine-pink flowers up to 9 cm long. It flowers from August to December.

Schlumbergera truncata (Haw.) Moran (216), a better known species, is often grown by people who usually grow other plants than cacti, and known widely as 'Christmas Cactus', or as 'Crab Cactus'. In older literature, it appears under the names *Zygocactus* or *Epiphyllum.* Unlike the other members of the genus, it does well on its own roots, providing it has a neutral to slightly acid compost. A mix of peat and loam with sand is suitable. In autumn and winter, it is best left in one place as changes in warmth and sudden movement cause it to drop its buds. Many hybrids have been cultivated from the original botanical species, producing flowers from

214

215

216

217

carmine-pink to pure white. *S. orssichiana, S. opuntioides, S. russelliana,* and some delicate hybrids must be grafted on taller trunks of *Pereskia, Pereskiopsis, Opuntia,* or *Eriocereus.*

Rhipsalidopsis rosea (Lag.) Britt. et Rose (217) has bright green segments similar to *Schlumbergera.* It differs from that genus in having square fruits and symmetrical, radial flowers. It comes from southern Brazil like the better known *R. gaertneri.* It can also be found under *Epiphyllum, Epiphyllopsis,* and it is widely known as 'Easter Cactus', because it bears flowers in April. Both species, as well as their cultivars, can be grafted to obtain more flowers.

Rhipsalis heptagona Rauh et Backb. (218). *Rhipsalis* is the largest genus of epiphytic cacti, comprising over 60 species ranging from Mexico to Argentina. Several species have been found in east Africa, in Malagasy (Madagascar) and Ceylon. They differ from the American species in the number of chromosomes, and it is probable that they were carried to their new homes by seed-eating birds. *Rhipsalis* species are hanging epiphytes, usually of cluster or shrub-like shapes, with cylindrical stems and tiny flowers. In some, the stems are flattened as in the related *Disocactus.* Cultivation also requires tropical conditions, particularly regarding humidity, warmth and an acid, humus-rich compost.

218

219

Chapter 4 THE CENTURY AGAVE

The agaves (Agavaceae) are a family of monocotyledonous plants related to lilies (Liliaceae) and amaryllis (Amaryllidaceae). Most species occur in dry or semi-arid areas, and they have succulent, water storing leaves, for example *Agave* and *Sansevieria,* or water retaining, more or less thickened woody stems *(Nolina, Dasylirion, Calibanus).* The Agavaceae include some sclerophilous xerophytes, that is plants with tough, leathery leaves and fibrous stems, which do not store water in water-retaining tissues but nevertheless withstand aridity extremely well, for example *Yucca.* All the agaves have leaves arranged in rosettes and flowers in panicles or clusters. In most genera, the flowers appear after many years, when the plants have accummulated sufficient food reserves. In *Agave,* this takes several decades (especially in cultivation). This fact gave rise to the legend that they, and sometimes even cacti, flower once in a hundred years (century plants). This myth concerns agaves and not cacti, although agaves will probably always be regarded as cacti by laymen, and it is not true. All agaves reach the flowering stage much earlier, and they have nothing in common with cacti except occurring also in the New World and prospering in a dry environment.

As a family, the Agavaceae were excluded from the Liliaceae and Amaryllidaceae by Hutchinson, in 1934; this classification is not found in older literature.

Mexico is considered to be the centre of distribution of the genus *Agave,* although many species grow in the adjacent areas of Central America, USA, and in the Caribbean islands. Over 100 species have been described. Closely related genera, sometimes classified under *Agave,* are *Furcraea, Manfreda* and *Hesperaloe.* In the wild, some species, such as *A. lechuguilla, A. stricta* and *A. horrida,* form extensive communities, *matorral,* dominating the rest of the vegetation.

Agave cochlearis Jacobi (219) is probably the largest species. It comes from the high mountains of southern Mexico, from rather humid localities at altitudes of up to 3,000 m.

220

Agave horrida Lem. (220, 228) originates from central Mexico. It can be found on sterile lava rocks, *pedregales,* north of the town of Cuernavaca. The leaf rosettes reach 1—1.5 m in height and the leaves have strong, sharp teeth on the edges. The yellow flowers are arranged in dense, cylindrical spikes. In cultivation these agaves are sensitive to the amount of moisture in their compost and they do not tolerate humidity at temperatures below 12° C. A lot of sunshine and a place in open air in summer will enhance the development and colouration of leaves, as for example in var. *gilbeyi* (228). *A. pumila* is the smallest species of the genus, reaching barely 10 cm in height over a period of more than ten years. Larger species need a longer period to accumulate enough storage materials, and only then will they mature and flower. The flowers are their first and last, because the leaf rosettes put all their strength into seeds and wither after the fruiting stage. Some species may form side-shoots and new rosettes before maturing. Some, however, do not produce side-shoots, for example *A. victoriae-reginae, A. utahense* and *A. purpusorum.*

221

Distribution of the genus *Agave* (including *Manfreda*)

222

Agave americana Marginata' Trel. (221) is a cultivar of the best-known agave species, *A. americana,* from central Mexico. The thickened strap-shaped leaves, toothed on the margins, grow rather smaller than the wild type. *A. americana,* together with many other species, is a plant which has been utilized by man since ancient times. The pre-Columbian Indian cultures in Mexico used the tough fibres from leaves for weaving, and for the Otomí tribes from central Mexico, the agaves were an important source of food — the so-called *agua miel* (honey water). When the inflorescences in *Agave* begin to form, many nourishing organic substances are concentrated in the leaf rosette, particularly sugars and proteins which are essential for the production of the flowers. By cutting out the flower stalk and drawing the liquid, the inhabitants obtained not only a nourishing drink, but also what is even today a pleasant beverage. After fermentation, the watery juice turns into *pulque* — a refreshing, slightly alcoholic beverage reminding one of young wine. Other species, such as *A. tequilana,* serve as a raw material for the production of the famous distilled spirits *tequila* or *mezcal.* The agave is also decorative. Many species are used for decorative cultivation in tropical and subtropical regions, mainly in the dry American areas. It would seem that in Europe they must be confined to the glasshouses of large botanic gardens or to parks and gardens in summer, but in fact many small and slow-growing species can enrich growers' collections. They are quite demanding, however, and their successful cultivation demonstrates the skill of the grower.

223

224

Agave potatorum var. **verschaffeltii** Bgr. (222) has a beautifully white-blue frosted epidermis and decoratively toothed leaves with black spines. It is a large species, but slow-growing in cultivation, and ten year old seedlings are less than 30 cm in diameter. As with most members of the family, the best way of propagation is from the seed. Viability, however, decreases fast in seeds with flat achenes. Most seedlings require permanent humidity in their first few months, a nourishing humus-rich compost and indirect sunshine. At the 2—3 leaf stage, the plant can be situated in a permanent spot in a greenhouse in full sunlight. Most species are not too particular about soil, and they tolerate higher and regular watering. The desert species from Baja California and northern Mexico are more sensitive.

Agave stricta Salm. (223) is more demanding in cultivation, but it is one of the most beautiful species. It is native to central Mexico, forming continuous carpets in many places. The seemingly tender cluster of several hundred bright green leaves looks innocent, but the needle-sharp tips warn that one has to move carefully. In cultivation, it should be situated in larger glasshouses, where the rosettes can be safely sited away from accidental touching.

Agave purpusorum Bgr. (224) is a representative of the wide-leaved species from southern Mexico, growing at higher altitudes in sunny sites. The leaves are approximately 75 cm long and the flower panicle is 3 m tall. *A. karwinskiana* is another interesting species with rosettes arranged on an elon-

225

226

gated axis, reminiscent of some species of the genus *Yucca.* Both species are highly decorative in glasshouse cultivation. They require a lot of sunshine and warmth, and dry conditions in winter.

Dasylirion glaucophyllum (225) see text on page 203.

Agave victoriae-reginae T. Moore (226). We continue the genus Agave with examples of the most beautiful but difficult plants. This species comes from northern Mexico, from limestone rocks and screes. The leaf rosettes take a long time to grow, and this succulent can function as an ornamental plant even in the restricted conditions of a greenhouse as can the following species.

Agave utahensis Engelm. (227) comes from Utah, northern Arizona and Colorado, where it grows in very arid conditions. It forms low rosettes of slender, bluish-green leaves terminated by long, sharp, dark coloured tips. It is a miniature species, 12—30 cm tall. Its cultivation is rather more difficult. Other handsome relatives are *A. parryi, A. chrysantha,* and *A. bracteosa. A. attenuata* from central and eastern Mexico is an ornamental garden or park plant. It has bluish-white, frosted leaves, mostly lacking the marginal teeth. Many small species are rewarding in collections, for example *A. geminiflora* with thin, supple leaves having a white, fringed epidermis in their lower third.

Dasylirion glaucophyllum Hook (225). All the 15 or so species of the genus have woody trunks and stiff, hard, toothed leaves. In *D. glaucophyllum,* they are bluish-green, in the species *D. longissimum, D. acrotrichum,* etc. they are bright green and elegant looking. Many species are therefore used as ornamental plants, mainly in Mexico. They can easily be grown outside their homeland, in glasshouses, halls, etc. They are rewarding modest plants, requiring a porous compost and over-wintering with occasional watering. Propagation is from seed.

Agave horrida (228) see text on page 198.

227

228

229

Yucca filifera Chabaud (229). The genus *Yucca* has about 20 species. They are long-lived plants with stiff leaves and usually a woody, more or less branched stem. They are distributed in the USA and in northern and central Mexico. Some species, such as *Y. carnerosana, Y. periculosa* or *Y. dactylis* are dominant landscape features. There is also the minute *Y. endlichiana,* reaching merely 30 cm in height, but developing an underground rhizome, similar to that of the best-known North American species *Y. filamentosa,* which can be cultivated outdoors in European conditions. The Mexican species are otherwise exclusively greenhouse or house plants. They are highly suitable for large halls and staircases. In summer they can be placed in gardens or parks, like *Dasylirion* and *Nolina.* They are not demanding and tolerate inexpert cultivation.

Nolina (Beaucarnea) gracilis Lem (230) is a plant of similar qualities to *Yucca.* At two to three years, its tender, bright green leaves are already very decorative. *Nolina* with its approximately ten species is distributed through central Mexico. *N. gracilis* comes from the Tehuacán region in the state of Puebla, and the similar *N. stricta* is its northern relative. In cultivation, they are easily propagated from seed. The seedlings require a loamy soil and tolerate ample watering in the first summer season. The fleshy

230

231

232

stem becomes woody in later years. The seedlings are then resistant to long periods of drought. In winter months, they tolerate complete dryness (occasional watering is suitable in the juvenile state), and temperatures below 10° C.

Sansevieria is an exception in the family Agavaceae in coming from Africa and Asia. It is a small-sized plant with fleshy leaves and mostly underground rhizomes or creeping stems. Like other agaves, the leaves contain tough fibres, in some places used for textiles.

Sansevieria hahnii 'Variegata' Hort. (231) is a cultivar derived from the larger *S. trifasciata.* It has become a popular ornamental foliage plant, forming thick, low rosettes with yellow variegated leaves. There is also a green leaved form, producing abundant offsets and quickly forming large colonies. It is easy to cultivate, tolerating ample watering and dry periods. In winter, it requires a light position and a temperature of 12—15° C.

Sansevieria stuckyi Good. (232) is a small species, approximately 20—30 cm tall with marbled leaves, and producing many offsets. In cultivation, it is sensitive to over-watering and winter cold. *S. trifasciata* is a well-known species with erect, sword-like leaves. *S. cylindrica* with cylindrical, green leaves is sometimes encountered in collections, but it is rather large.

Calibanus hookerii Trel. (233) forms globose, fissured, tuberous 'trunks', growing at ground level and reaching 40—50 cm in diameter. The interior of the woody stems contains (particularly in the juvenile state) fleshy tissues. The thin grass-like leaves are arranged in rosettes, the whitish flowers grow on short, straight stalks and form clusters. The seeds, as in the related species of the genus *Nolina,* are spherical achenes provided with membranous wings. *C. hookerii* is distributed in central Mexico. It is frequent on limestone hills, where the perfectly camouflaged rough 'trunks' and thin leaves are hardly distinguishable among stones. The native Indians use the leaves for basket weaving, and the plant is known under a local name *sotol* as are some *Dasylirion* species, for example *D. longissimum. C. hookerii* is an interesting plant for glasshouse culture. It grows well in a warm and sunny place, in a well drained compost and requires abundant watering in summer.

233

234

Chapter 5 RELATIVES OF 'FRANGIPANI'

Throughout the tropical regions of the world, one can see large shrubs with strong branches and lanceolate leaves, flowering throughout the year. In Central America, this plant is called frangipani *(Plumeria).* It belongs to the Apocynaceae which includes many well-known members of tropical vegetation, there being in all about 1,500 species. Of the many non-succulent, ornamental members of this family are the climber *Allamanda,* and the European *Vinca major* and *Vinca minor,* the vigorous, evergreen periwinkles. Of importance for growers of succulents are three genera, one from America *(Plumeria)* and two from Africa and Asia *(Pachypodium* and *Adenium).* They are stem succulents forming thin or leathery leaves. In most genera, the stems later reach larger, shrub or tree dimensions. Their tissues contain milky latex. The flowers are pentamerous (in fives), white, yellow and red. The seeds of most species are light, with a downy pappus, assisting their distribution by the wind. The desert species (some *Pachypodium* and *Adenium* species) are relatively demanding in cultivation, other African species of the same genera, like *Plumeria,* tolerate generous watering. Growth in cultivation is usually fast, and the plants soon reach a large size.

Plumeria rubra L. (234). The whole genus is distributed through the dry forests of Central America, in the Caribbean islands and in the northern parts of South America. Cultivars have been raised, derived from the botanical species which give a succession of flowers almost all the year, from white, yellow and orange to shades of red and purple. In cultivation, they are recommended only for larger glasshouses, where they quickly form shrubs or small trees and flower throughout summer. They are also excellent plants for light, dry offices, public halls, etc. They tolerate winter temperatures of about 15° C and irregular watering in summer, although abundant and regular watering is preferable. The recommended propagation is by cuttings from adult flowering plants.

Plumeria alba L. (235) is a botanical species with snow-white flowers, seen less frequently than the cultivars. It comes from the Caribbean islands.

235

Distribution of the genus *Plumeria*

236

Adenium obesum Balf. (236) is often called the Desert Rose; this is the best-known species, distributed through east Africa. It forms thick succulent stems, wide at the base and terminated by a crown of slender branches, on which bright slender green leaves, reminiscent of oleander, grow in the humid season. The genus *Adenium* occurs in central, southern and west Africa, on the Arabian Peninsula and on the Island of Socotra. In cultivation, the plants produce magnificent flowers, but propagation is difficult and slow. Grafting on *Nerium oleander* has proved successful; beautiful flowers appear within several months. In this way, however, one can obtain only a part of the interest of the plant, not its most remarkable feature, the massive stem (caudex). The scions can later be allowed to root out, but the *Adenium* tissues do so slowly and reluctantly. It is better to use hormones to speed up the formation and grow it on its own roots. All the species of the genus *Adenium* are sensitive to low temperatures. Watering in the rest period, when they have shed their leaves, should be avoided.

Adenium swazicum Stapf. (238). Experts have problems distinguishing the species of this genus as there are usually only minute differences in vegetative features. The flowers of all the species are very similar, they are pink and the margins of the inflorescences are darker.

Pachypodium brevicaule Bak. (237) is the smallest member of the genus which comprises some 20 species. The genus *Pachypodium* differs from the genus *Adenium* in the presence of spines on the surface of the stem, and in the seed structure; the hairy outgrowths are on one end of the seed only. The area of distribution covers south-western Africa and Malagasy. *P. brevicaule* is a rare, but also the most beautiful miniature representative of the Malagasian species. It comes from acidic siliceous rocks. It is propagated only from seed. The seedlings grow slowly. Other species are more vigorous and faster-growing. Amongst these is the following species.

237

Pachypodium lamerei Drake (the two plants on the left) and **P. geayi** Cost. et Bois. (the one with narrow long leaves) (239). They are also native to Malagasy. *P. lamerei* can be used as stock for grafting slow-growing, delicate species such as *P. brevicaule. P. geayi* is a beautiful greenhouse decorator thanks to its greyish frosting and long spines. Neither species flowers until it reaches over 1 m in height which means many years of cultivation. *P. lamerei* is the easiest one to cultivate. It tolerates abundant watering and loamy soil. Like the other *Pachypodium* and *Adenium,* it will not withstand temperatures below 12° C, especially if they last several days. Another interesting species is *P. namaquanum* with a strong spiny stem terminated by a leaf rosette reminding one of a cabbage head. *P. lealii, P. baronii* and *P. rosulatum* are large shrubby plants. Their seedlings have smaller flowers and form decorative shrubs within two to five years when grafted on *P. lamerei.* Propagation of almost all the species of *Pachypodium* is possible in cultivation only from seed. The seeds germinate easily and in their first year the seedlings need a warm and permanently humid place, protected from direct sunlight. The strong fleshy stems and branches of *P. lamerei, P. geayi, P. namaquanum,* used as cuttings, will root poorly, and the soft tissues are easily infected. The thin terminal branches of the shrub species *(P. lealii, P. baronii, P. saundersii)* will root better, but the cuttings will not produce the thickened 'elephant's legs' or bottle-like shapes seen in seedlings.

238

Distribution of the genus *Pachypodium*

239

240

Chapter 6 CARRION FLOWERS AND WAX PLANTS

The Asclepiadaceae family comprises some 2 thousand, mostly tropical, species. Only a few of them can be called plants of arid areas, that is xerophytes or succulents. The centre of distribution of the succulent types is in Africa, mainly in the eastern and southern regions. Members of the Asclepiadaceae differ from those of the closely related Apocynaceae in the more advanced specialization in their flower formation. They are pollinated in a way similar to orchids. The pollen grains are gathered into pollinia. In many species, pollination seems to be possible only through the agency of a specific insect, morphologically fitting into the flower's interior, like a key into a lock. This is the reason why the flowers of these plants are usually not fragrant, but unpleasantly smelling, for they are mostly pollinated by flies, attracted by the smell and laying their eggs in the flowers. This family includes the clambering lianes with succulent leaves *(Hoya),* or stout storage tubers *(Ceropegia, Fockea* and *Marsdenia),* and shrubs having thickened stems and thin branches *(Raphionacme),* low stem-succulents *(Caralluma, Stapelia),* and miniature, rare and delicate genera *(Whitesloanea* and *Pseudolithops)* very like cacti in appearance and native to the arid areas of east Africa.

Duvalia sulcata N. E. Br. (240). The genus *Duvalia* comprises small, creeping plants with cluster-forming, sometimes ribbed, angular or cylindrical stems, mostly with a tuberculate surface structure. They occur in west, south and east Africa, some species extending to Socotra and southern Arabia, to Yemen. The flowers are either single, or grow in twos or threes, as they do in *D. sulcata.* They are strikingly ornamental and colourful, and throughout summer they appear in succession. *D. sulcata* has the largest flowers (about 5 cm across), and their surface is grooved.

Ceropegia is a numerous and important genus, including both thin, creeping liane succulents, and stem-succulents with reduced leaves. They occur mostly in tropical, west, south and east Africa, extending to Asia.

241

242

Ceropegia distincta N. E. Br. (241) is a thin, twining plant from dry forests and bushland of southeastern Africa. The bright green stems form thin, barely succulent, lanceolate leaves in the humid season. The green stem is also important for photosynthesis. The flowers appear throughout summer until autumn. Together with the similar *C. haygarthii* and *C. sandersonii*, it is an easy and interesting flowering plant, recommended for collections. It can be used to cover the walls in greenhouses. All the species require a porous compost. Occasional, sparing watering in winter with temperatures about 12—15° C and abundant watering in summer are necessary.

Ceropegia stapeliiformis Haw. (242) is an interesting creeping species with thin stems, often several metres long. The new shoots take root and the old parts wither away. The leaves are reduced, hardly more than a few millimetres in length, the stems are succulent. The flowers are not freely borne, but they are beautifully shaped and coloured. The species is native to South Africa; its habitat is the bush.

Ceropegia juncea Roxb. (243) is a climbing species with stout tuberous roots and a thin, clambering stem. The small succulent leaves appear only during the humid growing season. In the dry months, their function is taken over by the

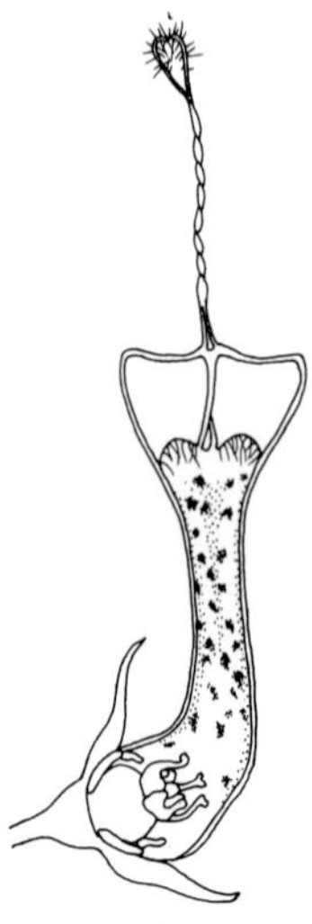

Cross-section of a *Ceropegia* flower. The pollinating insect makes an intricate journey to the bottom of the flower, to reach the stigma and anthers.

243

244

245

246

247

green stem. *C. juncea* comes from India, and in some places it is used as food by the native population. *Ceropegia* species from the Canary Islands, *C. fusca* with the grey-white frosted branches and *C. dichotoma* with green stems are an interesting group. Both form thick shrubs of erect stems and flower readily. In greenhouses, they soon reach larger dimensions and have to be reduced by judicious pruning. The cuttings can be used for propagation. Winter temperatures for the two species should not drop below 12° C when a completely dry place is recommended.

Tubers of *Ceropegia woodii* are suitable stocks for difficult species of the family Ascepiadaceae.

248

Ceropegia haygarthii N. E. Br. (244) is a species similar to *C. distincta.* It comes from South Africa. It can be cultivated in hanging pots or on windowsills. The flowers are pale pink and violet dotted. A well-known species is *C. woodii* appreciated by all the growers of the delicate species of this family, because the tubers can be used as stock for grafting (259). Cultivation is simple: the plant needs a porous rooting medium and abundant watering in summer. A cool and dry environment in winter is best.

Huernia confusa Phillips (245) is a small species with angular stems of bright green colour. It quickly produces clumps and flowers freely in various shades, equalling the ornamental *H. zebrina.* Both species come from the Transvaal in South Africa.

Huernia pilansii N. E. Br. (246) has tuberculate stems and reaches only small dimensions, 5–6 cm in height at the most. It is frequent in cultivation. The colour of the epidermis ranges from bright green to yellow-brown and russet and when exposed to sunshine, the plant turns darker coloured. The flowers appear in late summer and autumn and are dark brown with pale yellow dots. It is native to the Cape Province, occurring in dry bushland in shady places in the undergrowth and grass. Cultivation is easy if a dry, lightly shaded overwintering place is provided with a temperature above 10° C.

249

Huernia zebrina N. E. Br. (247). The genus *Huernia* comprises some 100 species. It resembles the genus *Caralluma,* but the flowers have a more sophisticated structure: they are mostly borne singly, but two to three flowers may sometimes be borne on one stalk. *H. zebrina* is the most popular representative in cultivation and is easy to grow. It requires humus-rich, porous soil, sufficient humidity and a careful, occasional watering in summer, and a light place in winter. Many species are particularly sensitive as regards their winter position which has to be relatively warm (about 14° C) and light. The humus-rich compost can, if it gets too moist, often cause infection or rotting. It is recommended therefore to take preventive measures by applying fungicides (Captan, Benlate, Dexon, etc.) in autumn, especially for difficult species. The infection makes the stems turn black and soft at the base, or may start at the top. The plant can be saved if transferred to a temperature of 25° C, with the affected tissues cut away, and the wounds dried with a fungicide. For these reasons, the delicate species should be watered only when necessary. The soil mix should be composed only of well-matured loam and sand, sterilized before use.

The genus *Caralluma* includes small plants with succulent, creeping and climbing, usually angular or cylindrical stems which are rarely tuberculate. The flowers grow at the ends of the erect branches, terminally or laterally, gathered most often into clusters, or forming large, simultaneously blooming inflorescences *(C. retrospiciens).* They are distributed from southern Europe to South Africa, but they also grow in Arabia, Socotra and Malagasy.

Caralluma europaea (Guss.) N. E. Br. (248) is the only species of the succulent Asclepiadaceae native to the European continent, and occurring also in north Africa. In the juvenile stage the quadrangular stems develop tiny leaves which later wither and fall off. It has profuse, tiny striped flowers appearing on the plant almost all summer. Cultivation is easy. A porous, humous rich compost and regular watering are beneficial.

Caralluma ramosa (Mass.) N. E. Br. (249) comes from South Africa where it forms large clumps of much branched stems. The stems are green, in the sun covered with a grey, wax-like coating, and the flowers emerge in late summer. They have an interesting colouration: a brown-purple background with yellow striping. Many other species are equally attractive. There are species with stems up to 1 m in length *(C. retrospiciens* and *C. penicillata),* with fine tufts of hairs on the flowers which move jerkily for a long while when shaken by a breeze.

Fockea crispa (Jacq.) Schum. (250) is a stem succulent with a thick woody and fleshy base (caudex) and relatively thin and twining branches. Some five species are known, coming from southwestern and southeastern Africa, where they inhabit steppe communities and deciduous dry forests. *F. crispa* is the best-known species, but it is rarely encountered in cultivation. It is attractive mostly at a later age, when the wrinkled base of the trunk can be used as a succulent bonsai. Propagation by cuttings is possible, but the caudex does not develop in this case. Cultivation is easy, with regular watering in summer. In winter, when the plant sheds leaves, a completely dry environment and temperatures above 10° C are necessary. The flowers appear in high summer and last until autumn. They have a relatively inconspicuous greenish colour.

The growers of indoor plants are familiar with the genus *Hoya,* the wax plant. The beautiful waxy flowers are not fetid but have a pleasant scent. Several dozens of species come from the dry tropical forests of India, China, Indonesia and Australia. They often climb like small lianes over the surrounding vegetation, covering bare rocks and slopes, droop from hanging rocks or dead trunks, or form shrubs. Not all the species are succulents. They are cultivated more frequently as house plants than in succulent greenhouses. Apart from *H. carnosa* and its cultivars, which can stand winter temperatures down to 5° C, hoyas need to be kept above 14° C, below this there is a risk of damage. In the summer months, they require abundant watering, a humus-rich loamy soil and a place in the sun, if they are to flower freely.

250

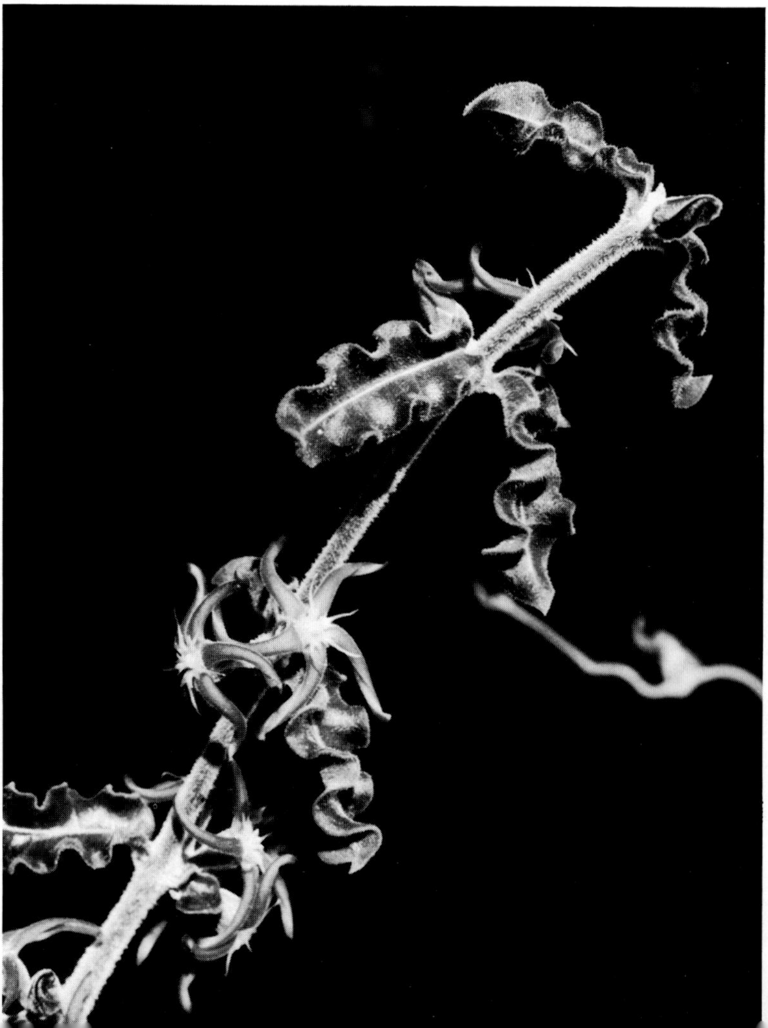

Hoya bella Hook. (251) is a shrubby species with erect main branches and drooping lateral ones. The leaves are slightly succulent, with a glossy green surface, and they are about 2—4 cm long. The flowers arise in summer, they have a porcelain, fleshy, waxy appearance and a pleasant scent. Reputedly, after grafting on the more vigorous *H. carnosa, H. bella* quickly forms shrubs profusely flowering throughout summer.

Hoya carnosa L. 'Variegata' (252) is a cultivar with succulent waxy leaves. The white margin reduces the plant's efficiency, slowing down growth and reducing the number of buds. Its chief ornament is, however, the leaf pattern, so it does not matter. The natural forms and cultivars of *H. carnosa* are valuable particularly in large rooms where they are very striking and also on staircases, in porches, etc. They flourish outdoors in sunny, protected places in summer months. The species is not particu-

251

252

lar about soil, it tolerates abundant watering and much lower temperatures than the other species which are important only in warm glasshouses. Most eventually grow to a large size except for *H. longifolia* and *H. linearis*, with tender elongated leaves and drooping stems. The tender flowers in both these species are similar to those of *H. bella*.

Piaranthus pulcher N. E. Br. (253). The whole genus comprises some 17 very small and rather similar species with segmented stems. The branches are angular to cylindrical, mostly green or brown-green. The species are native to South Africa, notably the Cape Province. They live in dense clustering colonies, in semi-shade or under the protection of the surrounding vegetation, only rarely in open country. In cultivation, they require conditions similar to *Huernia*, *Caralluma*, and *Stapelia*, that is a porous, rich compost, abundant watering in summer on warm days and careful handling in winter, when a position in a light place at 10—14° C with occasional sparing watering is beneficial. There is danger of infection by fungi, making the stems grow black and wither. *P. pulcher* is most frequently found in collections. The small bodies, about 15—25 mm across, are green to grey-green. The 10—12 mm wide flowers appear in late summer; they have a yellow background colour with red-brown spots and brush-like hairs. Other interesting flowering species are *P. punctatus, P. foetidus* and *P. decorus.*

Echidnopsis dammanniana Spr. (254) is a member of the genus which has elongated, erect or creeping stems, which often root, and tiny flowers. All approximately 15 species are native to southern Arabia, eastern dry areas of Africa, the Cape Province of South Africa and the dry coastal regions of west Africa, where they live in similar conditions to *Caralluma, Huernia* and *Piaranthus.* The flowers are brown-red, and yellowish-flowered species are also known. Cultivation is not difficult. Propagation is carried out in spring or summer, simply by cutting the stems which sometimes grow roots as they lie on the ground. Cuttings can be left to dry off for 4—6 days and rooted in sand.

Stapelia grandiflora Mass. (255) is perhaps the best-known member of a genus which comprises some 110 species. Unfortunately only a few of them are seen in collections. In appearance, the plants resemble the creeping and climbing stems of *Caralluma, Huernia* and *Duvalia,* but they are generally

253

254

255

bigger plants and have larger and more interesting flowers. *S. variegata* and *S. hirsuta* can be recommended for windowsill cultivation. They are as vigorous as *S. grandiflora.* Their unpleasant feature is the strong fetid smell of the flowers which attract flies, and their eggs can often be found in the blossoms on the second and third day they are open.

Tavaresia grandiflora (K. Schum.) Bgr. (256). *Tavaresia* is an interesting genus, comprising plants with tuberculate bodies and conspicuous, almost campanulate flowers. In *T. grandiflora,* the flowers are up to 14 cm long and about 5 cm across.

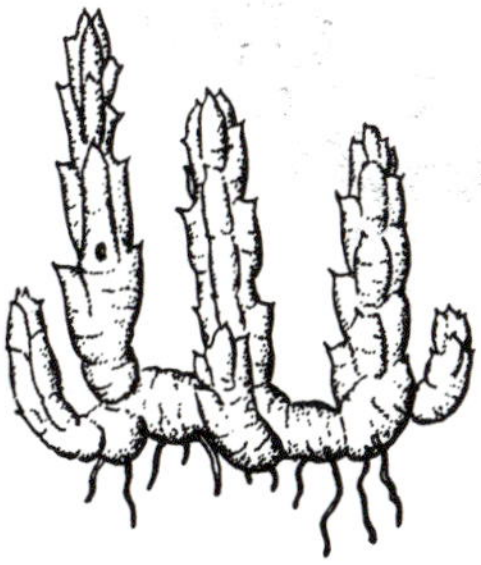

Prostrate stems of many species of *Caralluma, Hoodia, Stapelia,* etc. continue growing, put out roots and their older parts die off.

256

257

258

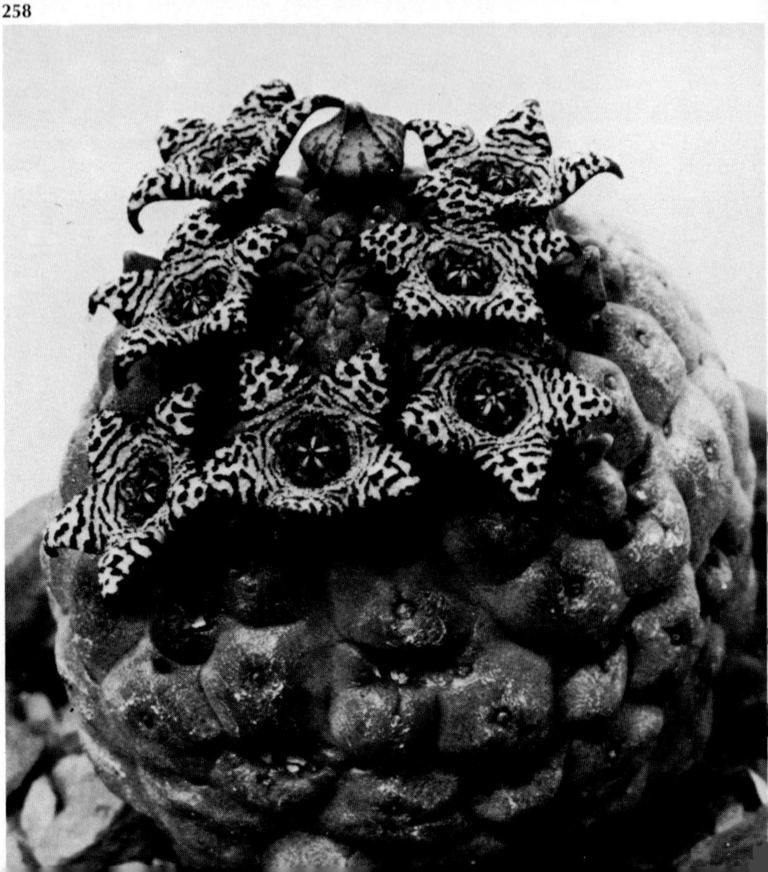

Distribution of the genus *Stapelia*

Tavaresia barklyi (Dyer) N. E. Br. (257) is a smaller relative, reaching only half the size. The stems are 8—12 cm tall. Both species are sensitive to over-wet compost and to drops of temperatures, and they flourish best when grafted on tubers of *Ceropegia woodii.* Even when this is so, winter temperatures must be kept at 15° C at least.

Trichocaulon cactiforme N. E. Br. (258). A genus with about 30 extremely interesting, highly succulent species. The stems are spherical to cylindrical, tuberculate, bearing small flowers. They come from south, southeastern and east Africa, including Malagasy. Cultivation is more demanding: careful watering and a warm place in winter are required.

Raphionacme spec. (259). This genus represents the so-called caudiciform types of the Asclepiadaceae, together with *Fockea,* and some species of *Ceropegia.* The caudiciform types are characterized by the heavy thickened base of the stem producing only slightly succulent branches with thin leaves. The leaves persist only in the wet season and fall off in the dry season. It is a rare succulent in cultivation. Propagation is from seed; cuttings give rise to exuberant non-succulent branches and leaves, but the characteristic caudex is not formed.

259

260

Chapter 7 SUCCULENT GROUNDSELS

The representatives of the family Compositae, one of the largest in the plant kingdom, are widespread all over the world and include some succulent forms, both leaf and stem types. Botanists based this widely differentiated family on the flower structure. What is popularly regarded as a single bloom in dandelions, dahlias, chrysanthemums, as well as in succulent members of *Senecio* and *Othonna,* is in fact an inflorescence. It is composed of a large number of tiny, tubular, densely clustered flowers situated on a flat capitulum and protected by the involucre. Two types of flower can be encountered; the inside flowers may be tubular and known as disc florets, while the circumference is composed of ray florets. Either type of flower may be absent, leaving the head entirely composed of one sort of floret. The seeds of the succulent groundsels are equipped with hairy parachutes assisting their distribution by the wind. The largest number of succulents is found in the almost ubiquitous genus *Senecio.* In this genus, we can encounter semi-desert shrubs but also trees, although some species form only low leaf rosettes. The centre of distribution of these succulent species is South Africa. Some are found in India, Malagasy, the Canary Islands and Mexico.

Senecio stapeliiformis Phillips (260) comes from the western Cape Province, occurring on the eastern slopes of coastal mountains suffering from lack of rainfall for long months, followed by a wetter season, bringing abundant moisture in the southern summer and autumn. *S. stapeliiformis* is a tolerant species, flourishing not only in greenhouses but also on window-sills, in humus-rich, regularly watered soil. In winter, the dormant plant should be kept in cool conditions at about 10° C.

Senecio haworthii (Haw.) Sch. Bip. (261). While the preceding species attracts growers by the shape and colour of the stem (and by seasonal flowers), this species has attractive leaves. It reaches up to 30 cm in height and forms a small

Distribution of the genus *Othonna*

261

shrub. The rare flowers are small and yellowish. Overwintering is as for the species above.

Senecio scaposus DC (262) is a low, rosette-forming plant with leaves covered with dense white hairs rather like fine felt as in the previous species. *S. scaposus* is sensitive to too much water in the compost during the resting period.

Senecio ficoides (L.) Sch. Bip. (263) is a very attractive species reaching a height of 1 m. It forms thickets of leafy stems. The leaves have a frosted, bluish, waxy coating, which can be easily wiped off. Many other species are found in cultivation, for example *S. articulatus, S. gregori* and *S. kleinia.* Their leaves are soon shed and only an interesting, often marbled stem remains. Other species are conspicuous by their creeping or hanging mode of growth and persistent, round, often coloured leaves as in *S. herreianus* and *S. rowleyanus.* Other species from South Africa or Mexico form stout succulent stems; *S. praecox* and *S. fulgens* are less well known examples. They can be shaped to look like aged, stunted trees — the succulent bonsai.

262

263

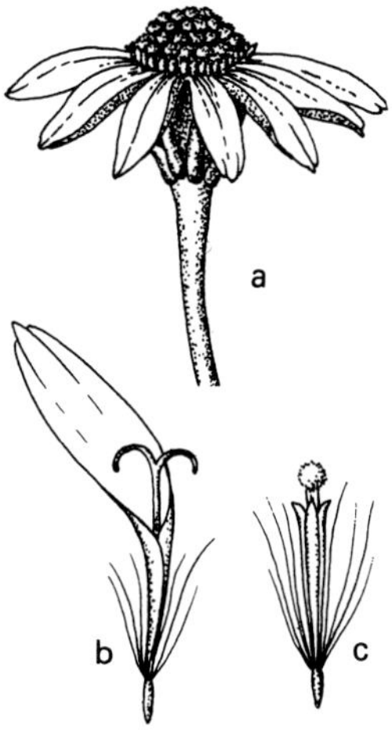

The apparent flower (a) in Compositae is a capitulum made up of many ray florets (b) and disc florets (c).

264

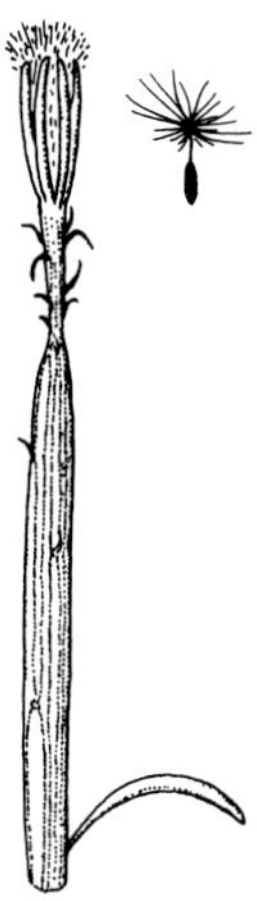

Fruit and seed in *Senecio*.

Othonna is another important genus of the family Compositae, and is confined to South Africa. Some 30 succulent species reach the proportions of small-sized shrubs, or have their stems hidden underground.

Othonna euphorbioides Hutch. (264) becomes larger in cultivation than in its homeland (where it measures only about 10 cm in height), and it later forms miniature, cactus-like trees. When the flowers wither, the stalks become woody and thorn-like. In cultivation, the leaves grow in autumn when watering is needed. The plants usually survive the summer heat without their leaves. Other species are mostly difficult to grow and are rare in cultivation.

265

Chapter 8 BROMELIADS

Bromeliads (Bromeliaceae) are the family of monocotyledonous plants which include the well-known pineapple. They can be small *(Tillandsia)* or robust plants *(Puya)*, with leaves arranged in rosettes. As with *Agave*, the inflorescence emerges after several years, in the centre of the leaves, exhausting the plant which perishes soon after the seeds are mature. In most cases (except *Puya*), a group of side rosettes is formed by offsetting before the flowering stage. The 2,000 or so species of the family are mainly forest dwellers; many are epiphytic, attached to tree tops, exposed rocky walls, etc. Only a few bromeliads can be regarded as xerophytes, or succulents. These plants can retain water in an interesting way: the leaves are covered with specialized scales, silvery-looking in dry conditions. They are made of cellular materials and are strongly hygroscopic. The scales absorb water from humid air, and the plants have the extraordinary capacity of using this physically retained water. This quality is perfectly developed in the so-called air plants (tillandsias), but is also found in other, non-epiphytic genera. It is disputable whether these plants should be included among succulents. Various studies have revealed similarities in their way of life and in many cases they have the special water-holding tissues in leaves as do many succulents. The atmospheric plants can be at least marginally added to the other succulents. Leaf succulence is most manifest in the terrestrial genera *Abromeitiella, Dyckia* and *Hechtia.*

Tillandsia ionantha (265) see text on page 239.

Abromeitiella brevifolia (Griseb.) Castell. (266). This genus comprises three species and comes from northern Argentina and adjacent Bolivia. At higher altitudes, the plants frequent slopes and are often found on bare rocks. Their hummock-forming growth enables them to retain water among their leaves for a long period. In addition, the top side of the leaves is developed into water-holding parenchyma, retaining water in the dry months. The

266

A wide leaf rosette *(Hechtia rosea)* is the typical growth formation in succulent bromeliads.

267

flowers of all species are tiny and greenish. In cultivation, the plants form dense hummocks. They are not too demanding and do well in an acid peat and sand mixture. In summer, they benefit from abundant watering and syringing with soft water. The winter temperature should not drop permanently below 10° C. At this time the plants require an absolutely dry environment.

Hechtia is a closely related species, including approximately 30 species from Mexico. The taxonomy of this group is very difficult and therefore as yet incomplete. The plants are terrestrial, with stiff, pointed, strongly toothed leaves, and a developped water retaining tissue on the upper side.

Hechtia argentea Bak. (267) comes from central Mexico where it grows on the vertical, inaccessible walls of deep canyons, as for example in the Barranca de Toliman. The surface of the leaves in this species is covered with silvery-white scales, and it can be classified among the semi-aerophytic bromeliads, partly obtaining water from the air.

Hechtia podantha Mez. (268). The plant is pictured in its native central Mexico, where it is found together with *Agave, Opuntia* and *Ferocactus.* It demonstrates the plant's extreme resistance to drought and scorching sunlight. It often forms large, circular, tangled thickets or clusters, which later wither from the middle.

Hechtia ghiesbreghtii Lem. (269). In the dry period, most species growing in the sun become red-coloured. *H. ghiesbreghtii* comes from central Mexico where it frequents mineral soils. Cultivation of this species and of its relatives is not difficult: a light, porous compost is beneficial, abundant watering in summer and full sunlight under glass if intense colouration is wanted. The flowers in all species emerge on racemes about 30—40 cm long but they are not very conspicuous.

The semi-arid regions of Brazil, Uruguay and Paraguay are the homeland of the closely related genus *Dyckia* with some 80 species. Their way of life resembles that of *Hechtia* in Mexico; they form large tight clusters, but the rosettes are smaller and less menacing-looking thanks to a finer toothing on leaves.

Dyckia marnier-lapostollei Rauh (270) is a very handsome silvery-white species, only recently described. It forms smaller clusters of relatively wide and soft rosettes. The leaves get a reddish to brownish shade in the sun. In cultivation, it requires relative warmth and humidity in winter.

268

Dyckia fosteriana L. B. Smith (271). Individual rosettes reach about 20 cm across and may form large clumps. Like the preceding species, it comes from southern Brazil, frequenting rocky cliffs and natural walls. It has narrow leaves with down-turned teeth, the whole covered with silvery scales, showing that the plant is partly storing moisture from the atmosphere. *D. fosteriana* and many other related species are excellent in cultivation. Many can be grown both in greenhouses and on window-sills together with hardier cacti. They tolerate irregular watering and temperatures down to about 10° C in winter. As with most succulents, they require a lot of light. The most vigorous types are *D. brevifolia, D. sulphurea,* and *D. rubra.*

The dyckias represent transitional types between terrestrial bromeliads drawing water and nutrients through taproots, and the second group, the so-called air-plant bromeliads, represented mostly by the

269

Distribution of the genera *Hechtia* (a) and *Dyckia* (b)

270

genus *Tillandsia.* It has a vast range of distribution, from Florida across Mexico and the Caribbean islands, Central and South America, to Argentina and Chile. Tillandsias can be found both in lowlands and at high altitudes in the mountains. Their existence depends on atmospheric humidity, that is upon water vapour in the air. Nevertheless, many of them occur in territories dominated by cacti and other succulents, often settling on them (276). Of great importance for tillandsias are their morphological adaptations; the thin grass-like leaves, folding horizontally in the dry season, and the coating of silvery-white cellulose scales, which in the dry state reflect the sun rays and maintain a humid microclimate within the epidermis, absorbing water when the atmospheric humidity increases. Another adaptation is their epiphytic life in tree tops, on the bodies of cacti or on rock walls. In the Peruvian

271

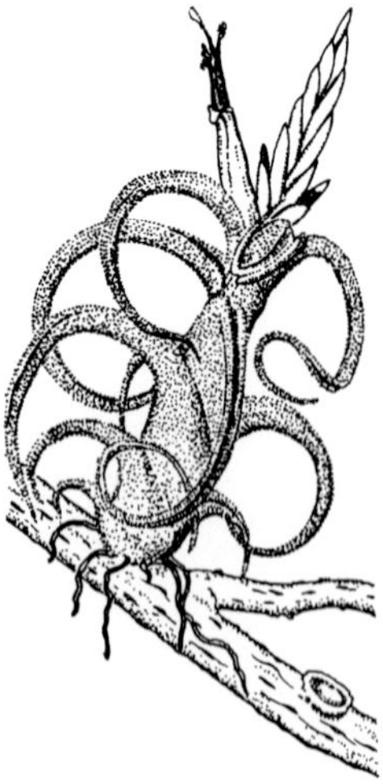

The aerophytic *Tillandsia* have a reduced root system; water and nutrients are supplied through the hygroscopic leaf surface.

272

273

mountains, tillandsias lie loosely on the sand of arid desert plains, where in the strong drafts of air their leaves can 'comb' the fog for droplets of water or use the night dew. In cultivation, the grower has to keep in mind these differences from other succulents and xerophytes. The xerophytic, atmospheric tillandsias will rather withstand dryness than constant moisture, and if the scales store up the absorbed water for too long and cannot dry up, they soon die. The quality of water is also important — xerophytic tillandsias need soft water, preferably rain water. In the summer months, they flourish if situated outside in tree tops, in semi-shade. They have to be taken indoors before the first frosts. They can stay all the year in cactus greenhouses, hung on pieces of bark or branches. They can be attached by nylon threads, or stuck by non-toxic, waterproof glues. They must never be fastened in moss clumps which can mould and infect the plants. The following species are recommended for cultivation in cactus greenhouses with winter temperatures of 10 to 15° C.

Tillandsia ionantha Planch (265) forms small rosettes with a bulbous base, the leaves grow up to 3—4 cm in length and turn red in the sun before the buds appear. The form in the picture is called 'Hazelnut', and it is the most beautiful and vigorous type for the dry air of greenhouses or homes with central heating. It comes from southern and southeastern Mexico, from semi-dry, bushy, sparse woods.

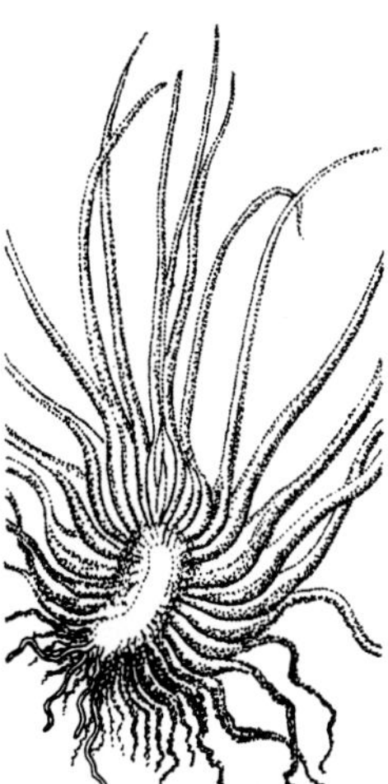

In *Tillandsia argentea*, the leaves grow on a thickened, bulbous and succulent base.

274

Tillandsia argentea Griseb. (272) is a very beautiful xerophytic species frequenting dry woods in southern and central Mexico. It has white to yellowish leaves with a bulbous succulent base, tapering into thin, hair-like tips. The flowers appear on a thin stem, approximately 10 cm long, and they are blue.

Tillandsia caput-medusae Morren (273). The leaf rosettes grow from succulent bulbous bases, in the wild often inhabited by ants. It is a suitable species for closed glass cases in the home or for greenhouses. It is native to Mexico and Central America.

Tillandsia brachyphylla Baker (274) has very decorative, silvery-frosted leaves. It comes from the Peruvian mountains. Its habitat is at high altitudes, and it is more demanding than many over humidity and a cooler all year position. In summer months, it does better outdoors in a garden than in a greenhouse.

Tillandsia regnellii Mez. (275) is similar to *T. brachyphylla.* Its native home is also on the eastern ridges of the Andes in South America. The two preceding tillandsias cannot withstand permanent humidity, and the plants have to dry off in moving air after every watering. Overheating under glass has to be avoided, because it can cause burning. Ventilation has to be efficient in the greenhouse.

Tillandsia recurvata L. (276) is a widespread species ranging from Florida to northern Argentina. It forms ball-shaped clumps of thin stems with grey, twisted leaves

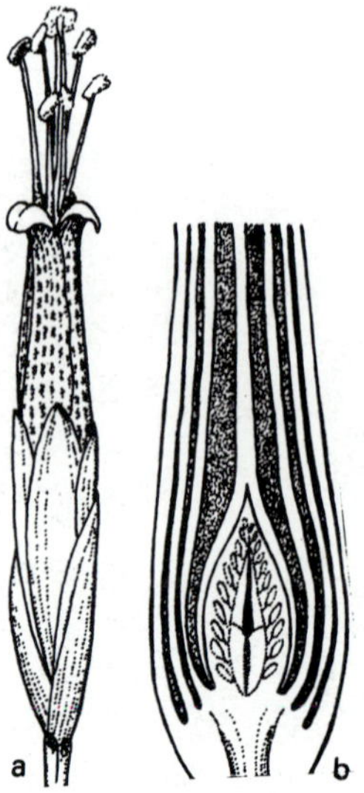

Flower of *Tillandsia* spec. (a) and its cross-section (b).

275

276

277

arranged in two rows. *T. recurvata* lives in the driest desert areas, often it can be seen even on electricity cables. It is one of the hardiest species in cultivation, tolerating a very dry position. In the picture it is growing on a columnar cactus.

Tillandsia atroviridipetala Matuda (277) comes from southern Mexico, from dry deciduous forests at higher altitudes. *T. plumosa* is a similar species but it comes from cool, humid mountain forests, and the leaves are covered with fine, snow-white hairs. *T. atroviridipetala* has green flowers and forms bulbous, succulent leaf bases. Tillandsias can be obtained from nurserymen, but their vegetative reproduction is not easy. Some of them *(T. ionantha, T. recurvata* and *T. schiedeana)* quickly form clumps of stems which can be divided and grown separately. Most other species, however, can be reproduced vegetatively only with difficulty. Propagation from seed is an important way of increasing stock but the cultivation of small seedlings is demanding. Sterile agar with mineral nutrients as for orchids, peat or sphagnum moss are used as a sowing medium. Such sowings must always be regarded as experimental and it is wise to consult specialist literature for advice.

278

Chapter 9 SUCCULENT ORCHIDS

The realm of succulent plants is extremely variable. We have explained that succulents are plants accumulating and retaining water, and since some of the orchids (Orchideaceae) have strongly fleshy pseudo-bulbs and leaves, there is no objection to their classification among succulents. Many orchids are found in areas affected for months by shortage of water. They occur for instance along the Pacific coast of Mexico, the homeland of many species of *Oncidium,* or at low elevations on the twisted trees of the dry coastal terraces of the Antilles. Here are found species of the genera *Broughtonia, Cattleyopsis* and *Laelia.* The Central American countries are the home of xerophilous epiphytes of the genera *Cattleya, Laelia, Brassavola, Chysis* and *Epidendrum.* Xerophilous and succulent orchids are also found in Australia *(Dendrobium),* Asia *(Vanda teres),* and in Africa *(Eulophia).* The genus *Vanilla* is widespread in the tropics and includes extremely hardy succulent species found in sunny areas. It would be paradoxical to leave out these succulent orchids. Their succulence is more conspicuous and their drought resistance much greater than in the forest cacti *Wittia* and *Disocactus!* Orchids are attractive thanks to their magnificent flowers, and their enormous interest is enhanced by their remarkable way of life and the difficulties of propagation which remained a mystery for a long time. The majority of them will continue to be a subject of interest only for the specialist orchid growers, others, however, can be cultivated together with succulents, particularly with the members of the family Bromeliaceae. Many grow best in small plant cases kept in windows or on balconies, others can be grown in a succulent garden greenhouse, some in the home. The majority of orchids thrive in warm conditions, and winter temperatures should not drop below 15° C for the tropical forest species, others are happy at 10° C and even below. The rest period should take place in a dry environment, but every other week, on a warm, sunny day, they should be sprayed with soft water. Succulent orchids are grown hung on pieces of bark or branches, usually fastened with wires or nylon. When planted, the base is wrapped in a mixture of fern fibre (*Osmunda* fibre) and moss *(Sphagnum).* Succulent orchids prefer a light, sunny position. They require regular watering and a temperature up to 30° C while they are growing though less than this is adequate. They need a great deal of ventilation. The following daily schedule is recommended: they are watered in the evening and the plant case or greenhouse is closed (to obtain high humidity), and opened before noon giving free ventilation for the rest of the day.

The genus *Oncidium* includes hundreds of species distributed throughout Central and South America, including the Caribbean islands. Most of them are denizens of humid forests, but several species occur in the dry coastal *chaparrals,* exposed to strong sunshine over winter when the only rare source of water is to be found in thin night mists or dew.

279

280

281

282

Oncidium maculatum Ldl. (278) comes from southern Mexico. It lives on stunted oaks or shrubby *chaparros (Curratella americana)* on the Pacific coast of the continent. The related *O. atropurpureum* is distributed on the eastern side of the continent, in the Antilles and Caribbean islands.

Oncidium cebolleta Sw. (279) is found in many states of Mexico and Central America. It is the hardiest species of the genus. The leaves are circular in cross-section. Other vigorous species include *O. cavendishianum, O. teres, O. splendidum, O. ascendens* and *O. stipitatum.* They do not tolerate a humid environment and require sunshine, warmth and a dry place. *Laelia, Schomburgkia, Cattleya* and *Brassavola* make up a beautiful group of succulent epiphytic orchids. Their hybrids are among the most popular marketed flowers.

Cattleya aurantiaca (Bateman) P. N. Don. (280). Although it has smaller flowers than many other species, it is an interesting and beautiful orchid, distributed through tropical Mexico and Central America.

Chysis bractescens Ldl. (281) is a member of a group of species with developed storage pseudo-bulbs. In the dry period, the leaves are shed. Many species inhabit Central and northern South America living in the dry deciduous forests. *Cyrtopodium* and *Catasetum* are similar genera.

Catasetum maculatum Kunth. (282) from Central America tolerates drought in winter and it is suitable for cultivation with succulents. Other suitable climbing succulent species are those of the genus *Vanilla.* The terrestrial species, spending the dry period leafless, are another interesting group including *Bletia* from America or *Eulophia* from Africa. To conclude, the succulent orchids are easier and more rewarding in amateur collections than many orthodox succulents like *Adansonia, Pachycormus,* or *Adenia,* which soon reach huge size but rarely large enough to bear flowers in cultivation.

Brassavola glauca Ldl. (283) is another interesting xerophilous epiphyte, encountered in Central America, and suitable for cultivation.

Laelia glauca often grows on bare branches, exposed to scorching sunlight.

283

284

Chapter 10 STONECROPS AND THEIR ALLIES

The Crassulaceae is one of the largest families of succulents, comprising many hundreds of species. Like Cactaceae and Ficoidaceae, it is a family with almost all its members having succulent adaptations. Unlike cacti, the succulence is mostly manifested by thickening of leaves, and only a few species have markedly succulent stems. The species with leaf succulence often have reduced stems, but they develop rosettes of leaves. Some species are shrub-like. Crassulas are mostly small-sized, often minute plants. Extreme dimensions are reached by *Crassula arborea,* but its 2 m height is by no means a record in the world of succulents. Most species are perennial, although annual or biennial species also exist. Members of the Crassulaceae have the amazing capacity of regenerating from tiny branch fragments, and in many species even from individual leaves. The structure of their flowers is also characteristic: they are in fives (pentamerous) and mostly small, often arranged in many-flowered inflorescences. There are few places in the world where Crassulaceae cannot be found — they are distributed from the tropics to the arctic and from low coastal deserts to mountain altitudes. Some species are frost-resistant and they are popular on rock gardens.

The genus *Aeonium,* comprising some 45 species, is centred in the Canary Islands with a few species on the Cape Verde Islands, in Morocco, Madeira, and on several Mediterranean islands. Almost all of the Canary Islands have their own specific species which occur only at certain altitudes and in particular ecological conditions. Some experts consider them analogous to Darwin's finches from the Galapagos Islands — and this insular specialization to specific living conditions is original and unique. All the species form either terrestrial rosettes, or bear leaf rosettes at the ends of woody branches.

Aeonium nobile Praeg. (284) is one of the largest species, its leaf rosettes reaching about 50 cm in diameter! They remain solitary even in old age. The leaves are tough and fleshy. Maturing is slow and flowers can be expected after seven to ten years. In the middle of summer, a stalk with several hundreds of tiny reddish blooms emerges from the centre of the rosette. It also

285

286

means that the rosette, exhausted after flowering and seeding (as with all other species) will die when the flowers wither. Propagation from seed is easy. Almost all aeoniums can be propagated from leaf cuttings: mature leaves are left to dry up for two to three days, and then placed on slightly humid peat and sand underneath a glass cover or a plastic bag. Keep at a temperature of about 25—30° C, providing a high humidity. The leaves will root within a month.

Aeonium arboreum var. **albovariegatum** (West) Boom (285). The botanical species *A. arboreum* is distributed in coastal areas of Morocco, Portugal and Spain, but it extends also far into the Mediterranean region, occurring on most of its islands. The colourful cultivars are decorative plants and their cultivation is easy. They tolerate regular watering, a nourishing compost and minimum winter temperatures of 6—10° C.

Aeonium arboreum var. **atropurpureum** (Nich.) Bgr. (286). The leaves of this cultivar are dark purple-red. They take on an intense colouration when in full sunlight.

Aeonium tabulaeforme (Haw.) Web. (287) has small succulent leaves, bright green and hairy on the margins. It does not produce offsets and can be propagated from seed or the leaves. It requires careful watering in cultivation, and

287

288

a temperature above 7° C in winter. It can reach up to 50 cm across if permanently situated in a shady site.

South Africa is the homeland of the interesting genus *Adromischus.* All the 40 or so species have extremely thick leaves. They grow in semi-arid areas, forming tiny shrubs and producing small inconspicuous flowers, arranged in a sparse spike.

Adromischus cristatus (Haw.) Lem. (288) is barely 10 cm tall with pale green leaves. As with other species, the leaves often form aerial roots, and the leaves if de-

289

tached will root readily. They often fall off themselves and can be used in this way.

Adromischus maculatus (Salm.) Lem. (289) has red spotted leaves, without the curly, blunt ends. Propagation is by easily detached leaves.

Adromischus cooperi (Bak.) Bgr. (290). The leaves on short stems are densely aggregated and the spots on the leaves are dark brown to black. These small shrubby plants are decorative when grown on window-sills. The requirements of all the species in cultivation are roughly the same as for those of *Aeonium* — winter temperatures can drop to 10° C.

Cotyledon is another important genus. Like *Adromischus,* it includes plants distributed exclusively in the Old World. Most come from South, south-eastern and east Africa and adjacent islands, while a few species are found on the Arabian Peninsula. In the wild, *Cotyledon* species can be found in various communities, mostly irregularly supplied with water. The many species include various forms of small deciduous shrubs to leafy, stunted trees of 0.5 m in height. The easiest propagation is from stem cuttings and in some species from leaf cuttings.

Cotyledon sinus-alexandri V. P. (291) is a tiny species with prostrate creeping stems, pale bark and non-caducous leaves, which are nearly spherical. Like the preceding species it flowers profusely. The flowers are a striking pink.

Cotyledon buchholziana Stephan et Schuldt (292) is a particular and demanding species reaching only small dimensions. The leaves are shed early, and only bare branches with colourful traces of the leaf bases remain. It is native to the arid regions of Namaqualand.

290

Distribution of the genus *Cotyledon*

291

292

293

294

Cotyledon wallichii Harw. (293) is a somewhat more robust species, with stems up to 35 cm tall, often with a whitish, waxy frosting. The relatively long, cylindrical leaves appear only during the growing period and fall off in cultivation when watering is stopped. The leaf bases survive, decorating the succulent stems like the segments of a pineapple.

Cotyledon paniculata L. f. (294) reaches almost tree-like dimensions growing to 2 m in height. The leaves are usually shed in the dry season, but older specimens remain impressive, giving the appearance of succulent bonsai. Cultivation is not easy as with the other above mentioned species, and it is recommended only for warm greenhouses. The greatest difficulty is its habit of coming into growth in the later autumn months, which are the least suitable because of the lack of sunshine. Since the plants must be watered at this time, the temperature in which the succulents are kept has to be increased to at least 20° C by day and 15° C at night.

The genus *Dudleya* comprises xerophilous plants from the westernmost state of Mexico, Baja California and adjacent southwestern USA. They form leafy rosettes, in many species covered with a white waxy frosting which repels water. In rain or during watering, the water runs down the leaves in large drops, but it washes away some of the bluish-white waxy coat. The 30 or so known species are not easy to keep successfully in cultivation, and their propagation is even more difficult. Propagation is only from seeds which are very fine like those of *Eche-*

295

Distribution of the genera *Dudleya* (a) and *Echeveria* (b)

296

veria. Cultivation of seedlings in the first weeks requires care and expe- rience. Propagation from detached leaves is almost impossible — the leaves will form a callus or roots, but not buds and new young plants In cultivation they require a place in the greenhouse in a very sunny spot, and winter temperatures should never drop below 12° C and the plants should be kept quite dry. In summer, watering must be done with care.

Dudleya farinosa (Ldl.) Britt. et Rose (295) has slender leaves but forms well-developed ro settes. It comes from Baja California, from the dry Pacific regions Cultivation is relatively easy as with the thermophilous echeverias such as *E. laui. D. farinosa* can be propagated by cutting off the side rosettes though this is not easy. Propagation from leaves has so far failed.

Dudleya pulverulenta (Nutt.) Britt. et Rose (296) is the most common species, ranging from southern California and Arizona in the USA to Sonora in Mexico. I forms rosettes composed of about 60–70 tongue-shaped, grey-white leaves. The flower stalk grows in summer, reaches up to 80 cm in height, and it bears many dozens of red to orange blooms, which unlike those of *Echeveria* are wide-open. The plants produce side-shoots which can be removed by carefully cutting free at the base and potting separately.

The South African genus *Crassula* gave its name to the whole family. The many dozens of species include miniature creepers or climbers as well as robust tree-like or shrubby plants. The flowers are not of great importance, and with a few exceptions, are inconspicuous in size and colour.

Crassula hemisphaerica Thunbg. (297) comes from semi-arid areas in southwestern Africa. Its pale green leaves are gathered in rosettes on short stalks. The flower shoots appear in spring and bear tiny whitish blooms. Cultivation is easy. The plants need a loam-based compost and respond to regular feeding. They also need humidity during the growing season, which usually extends to autumn months when temperatures above 15° C should be maintained. The rest period begins at the end of the year, watering must then be kept to a minimum, and if the plants are kept at temperatures around 10° C or lower, applications of water should be avoided. Another popular species is *C. (Rochea) falcata,* with grey, felted, flat leaves and a striking red-orange inflorescence. It is larger than *C. hemisphaerica.*

297

298

Crassula barbata Thunbg. (298) resembles *C. hemisphaerica;* it forms rosettes of fleshy flat leaves on short stalks. The margins bear conspicuous, thin, glass-white barbs. Its habitat is stony areas of the Cape Province of South Africa. Like many other species of this genus, it lives in semi-shade underneath bushes, or in open country exposed to sunlight. In cultivation it is easily propagated from detached leaves or stem cuttings.

299

Crassula perforata Thunbg. (299) is a prostrate to climbing species. The blue-green, often white-notched leaves are joined at the base. The name *perforata* corresponds to the fine dotted structure along the edges of leaves. Flowers can be expected only on older specimens. They are white, inconspicuous and hardly 1 cm across.

Crassula 'Morgan's Beauty' (300). As well as the botanical species, many hybrids have been developed in cultivation. These cultivars are generally easy to grow: they usually require higher humidity and regular feeding, tolerating higher temperatures all year long, so are better as house plants.

Crassula mesembryanthemopsis Dtr. (301) is a small, striking species with leaves arranged in rosettes, coming from sandy semi-deserts of southwestern Africa. The rosettes measure about 4—5 cm across and in older specimens the underground tuberous roots are prominent. The upper triangular leaves hardly show above the ground. The flowers appear regularly every autumn, sometimes at the end of summer.

Crassula lycopodioides Lam. (302) is a small shrubby succulent with erect, thin stems, tiny scaly leaves and inconspicuous whitish flowers, opening in late spring but rare in cultivation. It comes from southwestern Africa, forming large hummocks in bush-covered locations. Both by its appearance and its form of growth, it is like small species of club moss *(Lycopodium)*. The densely leaf-covered stems are rewarding in cultivation, easy to grow, and used for decorative purposes in the home.

Crassula pyramidalis Thunbg. (303). The erect stems often exceed 10 cm in height. The triangular leaves are densely arranged on the stem. Cultivation of this

300

301

302

303

304

species is relatively difficult. It requires a very porous, light potting mix and careful watering. The growing season is the late summer period or autumn. As with many other delicate members of the Crassulaceae, it can be grafted onto vigorous species of *Sedum*, or on *Kalanchoe tubiflora, K. daigremontiana,* etc.

Crassula arta Schoenl. (304) comes from the Cape Province in South Africa. The densely leafy, straight stems reach 10—15 cm at maximum, branching out at the base and forming dense groups. In the dry season, the leaves become pressed together and white frosted. Cultivation is similar for most of the species — very simple, requiring regular watering in the growing season. The plants tolerate an absolutely dry environment in the rest period, that is mainly in summer. Some of the species vegetate at this time, and the grower will, through experience, gain a knowledge of their needs when dormant or partly so.

Crassula arborescens (Mill.) Willd. (305) is probably the largest species of the genus, and a giant in the family thanks to its 2 m stems. The thick ellipsoid leaves and massive stems, easy cultivation and simple propagation make this species desirable even for the beginner and it is often seen on windowsills of the non-specialist. *C. arborescens* is one of the most rewarding and decorative succulent plants. It tolerates abundant watering even in winter (providing it is situated in a warm place). The small whitish

305

306

flowers are rarely produced and then usually in the very dry and hot environment of a greenhouse. *C. arborescens* is propagated not only from stem cuttings, but like all crassulas also from individual leaves which will quickly take root in a warm and moist environment and produce new plants.

Crassula marnierana Huber et Jacobs. (306) comes from southwestern Africa. It is a small species, densely covered with leaves and having tiny whitish flowers. As with the preceding species, it has flat inflorescences at the ends of stems. It is not difficult to grow, requiring a well-drained, sandy mix.

Crassula socialis Schoenl. (307) comes from the Cape Province where it frequents bare rocky terraces and crevices. In the wild, it usually forms low, prostrate clumps of short stems with leaves arranged in rosettes. It flowers profusely in early spring, producing white blooms. Cultivation is easy. The plant requires some moisture even in winter when buds are formed. It is successfully propagated by detached leafy stems, sometimes forming aerial roots.

307

308

309

Crassula cornuta Schoenl. et Bak. (308) is a very handsome miniature species, reaching barely 10—12 cm in height. The leaves are covered with grey-white waxy frosting which creates a decorative effect. The inconspicuous whitish leaves grow in winter. *C. cornuta* is also a native of southwestern Africa.

Crassula mesembryanthemoides D. Dietr. (309) forms in cultivation dense shrubs up to 30 cm tall. Its pointed leaves, covered with white, translucent hairs, are its main ornament. The whitish blooms are interesting, but rather spoil the appearance of the plant.

The genus *Crassula* deserves to occupy a more important place in cultivation. It comprises plants of many forms and most are rewarding and easily cultivated with the unique quality that the main growing season takes place in autumn and winter, so bearing flowers at the time when the other plants are resting and seem lifeless.

The New World is the homeland of two other genera of leaf succulents: *Echeveria* and *Graptopetalum.* The centre of their distribution is Mexico, although they can be encountered in Central America, in the northwestern states of South America, and in a few cases in the south of the USA. It is interesting that none of these genera reaches the Caribbean region and the Antilles. Morphologically, all the plants have leaves arranged in rosettes. A closely related genus is *Pachyphytum,* found only in Mexico, but its leaves grow on elongated stems and are differently arranged. The plants have a shrub-like appearance. The genus *Sedum* is also close systematically, many species occurring throughout Central America and chiefly in Mexico, though the genus as a whole is world-wide. The genera *Villadia* and *Tacitus* are related to the Mexican *Echeveria* and *Graptopetalum.*

310

311

Echeveria nodulosa (Bak.) Otto (310) seems to be distributed over a wide area of central Mexico, at higher altitudes up to 2,000 m. It is not difficult to find this species in the wild. Handsome specimens are found beneath trees or shrubs, in a mineral soil. The flowers are pinky-white and are very decorative. *E. longissima,* a similar species, has red-green flowers and lives in more humid conditions.

Echeveria elegans Rose (311). The native home of this blue-green leaved echeveria is in central Mexico, in the mountains north-east of the capital, Mexico City. It ascends to about 3,000 m, and can be found in pine woods, in clearings on rocky plateaux and rock-walls exposed to the sun. It is one of the hardiest species, often planted in parks for summer. The variety *hernandonis,* called after the contemporary expert on succulent plants Hernando Sánchez-Mejorada, has the most handsome compact rosettes. It forms dense cushion-like clumps of small bluish rosettes. The flowers appear in autumn or in winter. If the plants are placed in a cool and absolutely dry spot throughout the autumn and winter months, the flowers will be delayed until late spring.

Echeveria gibbiflora DC. (312). The form shown in the picture is not commonly cultivated. The var. *metallica* is better known. Together with other forms, it comes from central Mexico, where the plants frequent deciduous woods at altitudes around 2,000 m. It also grows on drier cliffs where there is more sunshine. *E. gibbiflora* is a massive species, often 1 m tall. The inflorescence reaches 50 to 60 cm in length and produces dozens of red blooms. It is a rewarding and hardy species, requiring a sunny spot. It is easily propagated from stem cuttings, using the new young growth which is made after the plants are pruned.

Echeveria agavoides Lem. (313) is about 10—12 cm tall with naked, bright green, waxy leaves. They are fleshy and tough, and in some forms, especially when growing in a very sunny place, the epidermis turns bronze-coloured or brown-red. *E. agavoides* comes from higher altitudes of central Mexico. It is easy to cultivate and is best kept during winter in an absolutely dry environment and at a low temperature. It is easily propagated from individual leaves. The red and yellow flowers grow early in spring. The interesting variety *corderoyi* has finer rosettes with denser leaves.

312

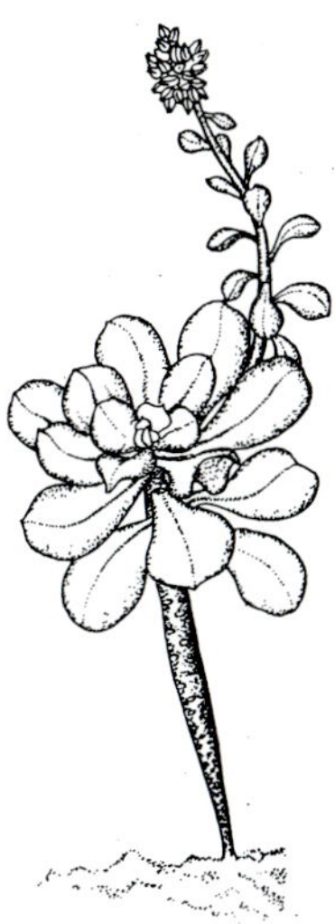

A dense leaf rosette is the most frequent growth formation in *Echeveria* (*E. multicaulis*).

313

314

Echeveria peacockii Croucher (314) is a small species, 6—7 cm across, with rosettes of blue-green, often grey-white frosted leaves. The edges and tips are red or red-brown, particularly when kept in a sunny place. It produces red to red-orange flowers in spring. Like the preceding species, it is easy to cultivate and deserves to be more widely grown both in collections and in the home. It is indigenous to the state of Puebla, coming from the area around the city of Tehuacán in Mexico. It is propagated from detached leaves.

Echeveria purpusorum Bgr. (315) is a small, ornamental species with spotted leaves. The flowers are yellow and appear in late spring. It is native to southern Mexico and its cultivation is more difficult than the preceding species. It requires a warm place all through the year, winter temperatures never below 14° C, and very little water in winter. It is propagated from seed which is extremely fine as are those of the whole genus. Seedlings require frequent watering and shade in the first weeks. Detached leaves can also be used, taking longer to root than do most species even in warmth and moisture. As with *E. agavoides, E. peacockii, E. glauca* and other small, rosetted species with thick leaves, *E. purpusorum* is a handsome ornamental plant, bringing welcome variety to a cactus collection. It is best to plant the rosettes in groups, when it looks rather like the European orpine *(Sedum telephium).*

315

Echeveria alpina E. Walth. (316). The 10 cm wide rosettes have pointed leaves which are blue-green to greyish. The flowers emerge in spring. The specific name *alpina* implies that the plant comes from high elevations. It is distribut-ed in the mountains east to Mexico City, up to almost 4,000 m. It is frequent in rather humid areas, covered with pine or oak-pine forests It grows on stony debris, on decaying rocks covered with moss and lichens. In spite of these difficult natural conditions, the species is vigorous and it tolerates low temperatures all the year (particularly in winter, when they can drop to 0° C). Full light is required in cultivation otherwise the stem elongates and the plant changes its appearance. It is easily propagated from leaves.

316

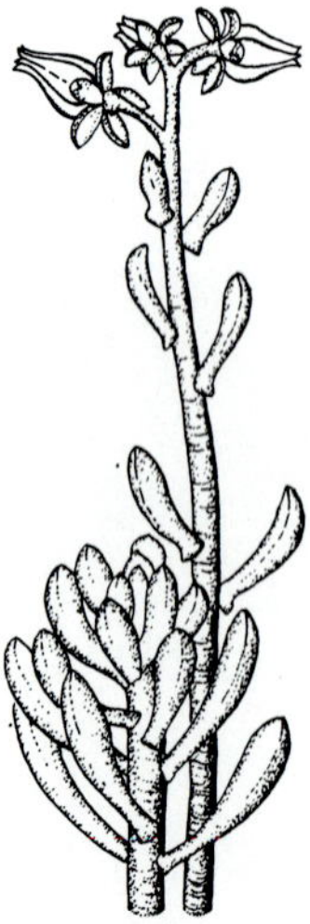

Other species of *Echeveria (E. nuda)* have leaves in alternate arrangement on an elongated stem.

317

Echeveria laui Moran et Meyran (317) was described only in 1975 and named after its discoverer A. B. Lau, a contemporary expert on the flora of Mexico and other American states. The plant comes from the beautiful region of central Mexico, around the Tomellin Canyon in Oaxaca, an area particularly attractive for succulent collectors. It is a hot region with summer temperatures above 30° C. *E. laui* is confined to a single narrow side-canyon at 500—600 m growing on steep, inaccessible rocks of red conglomerate. It grows only on the northern cliffs, shaded in the dry winter season. Entrance into the narrow pass is possible only through the bed of a seasonal stream, filled with water in summer. At that time, the place is inaccessible not only because of the stream, but also due to the high temperature and air humidity, making the climate unbearable. *E. laui* forms rosettes up to 25 cm across, especially in ideal conditions in cultivation. The rosettes are composed of more than a hundred white-blue frosted leaves which are tongue-shaped and fleshy. The flowering season in the wild is restricted to the beginning of the year (February, March), but delayed in cultivation to spring (April, May). The flower colour is orange, and it is extremely ornamental in combination with the blue-white frosting. Cultivation of *E. laui* requires fairly high temperatures throughout the year (above 14° C in winter), and careful watering. In winter, the plant must be kept almost dry, possibly watered carefully occasionally on warmer days to prevent over-desiccation. Propagation is difficult: if done from seed, the seedlings have to be carefully handled when they are young; they need an acid compost, shade, warmth, and constant humidity. Propagation from leaves also requires high temperatures and moisture. The choice of parent plant is also important: the leaves of adult specimens are slow to take root, while the leaves of small, juvenile plants will root readily, as well as the leaves from flower stalks. July and August are the best months for propagation.

Echeveria setosa Rose et Purp. (318) comes from the warm regions of southern and central Mexico. It occurs in light woodland, usually growing beneath the protection of taller vegetation. The leaves are covered with silvery-white hairs, preventing excessive evaporation. Since the plant's habitat is not too dry, regular watering is recommended in the summer growing season. It flowers profusely in spring.

318

319

Echeveria shaviana E. Walth. (319) is a forest species from the northern regions of Mexico. It frequents soil rich in humus, underneath pines and oaks, amply supplied with rainfall during summer and autumn. The dry winter season is cooler in the mountains, with frequent fog, dew, occasional rains or hoarfrost. Thanks to its bluish leaves, *E. shaviana* is very striking and is an easily cultivated species.

Echeveria gracilis Rose (320) comes from high altitudes (up to 3,000 m) in central Mexico. It grows in places where the winter temperatures drop to zero, and the dry vegetation is in the morning covered by hoarfrost. In spite of this, in cultivation the plant grows during the winter usually bearing flowers at this season too.

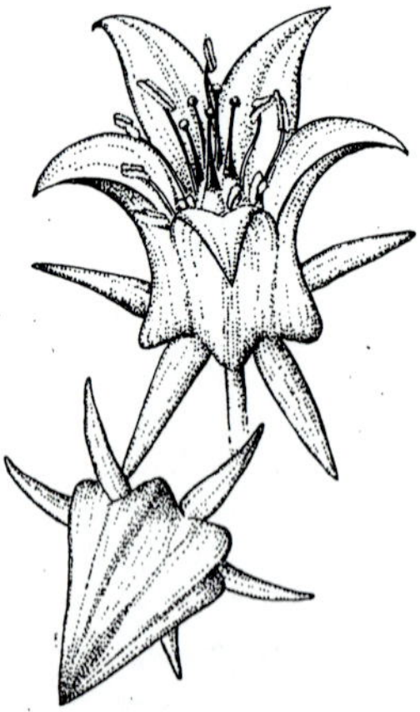

Flower and bud in *Echeveria*.

320

321

Echeveria dehrenbergii J. A. Purp. (321) hardly reaches 6 cm across. The bluish-green rosettes grow on limestone hills and rocks in southern Mexico. In cultivation, it is a rewarding, easily grown species, with well-marked red leaf edges. The flowering season is restricted to the short autumn and winter days, as in most other species. By maintaining low temperatures, the flowering period can be delayed until the early spring months. The flowers are striking, orange-red and persist for many weeks.

Cultivation of echeverias in European conditions is easy, being approximately the same as for *Crassula.* They flourish in a well-drained compost and respond well to feeding. In summer the plant puts on new growth. At this time regular and plentiful watering enhances their development and colouration, and they will form large rosettes. Many species in the wild are found in shady places, for example *E. nuda,* living epiphytically on oaks, but even these species should be grown in full sunlight in temperate conditions. Echeverias can be propagated either from seed, which is tiny, the seedlings requiring great care in the first few months, or from detached leaves. Leaves from mature or old plants are best not used for propagation as they root slowly. The tiny leaves of immature rosettes will take root much quicker and far more successfully. The small leaves on the flower stalk can also be used.

Mexico is also the homeland of the genus *Pachyphytum* comprising approximately 10 species. All are shrubs with fleshy stems and strongly succulent leaves. The flowers are arranged in spikes and the shape of the individual blooms is similar to that of *Echeveria* (the two genera are closely related and have produced many hybrids). Cultivation is the same as for *Echeveria.*

Pachyphytum oviferum J. A. Purp. (322) comes from northern Mexico and grows at altitudes of about 2,000 m. Nevertheless, it needs warmth and sunshine, and it should be grown in a sunny cactus greenhouse. Like *P. bracteosum* from central Mexico, *P. longifolium* and *P. viride,* it is an ornamental succulent, exciting admiration when grown well.

Another Mexican genus — *Graptopetalum* — is closely related to the genera *Echeveria* and *Dudleya.* It comprises small, rosette or shrub-like succulents, differing from the preceding species in their flowers, which are wide open, with petals grown together only at the base, and usually marked with darker dots.

Graptopetalum amethystinum (Rose) E. Walth. (323) is native to western Mexico where it lives on limestone rocks. Unlike the other species, the ovate, fleshy leaves are carried on long stems. The colour of the epidermis ranges from blue-green to violet, which gives it its specific name *amethystinum.* Cultivation is easy: the plant requires plenty of water in summer as well as warmth, and flowers at this time of the year.

Graptopetalum macdougallii Alexander (324). The rosette of blue-green, wax-frosted leaves reaches 6—7 cm in diameter. In winter (particularly in natural conditions) it will shrivel: most leaves dry up and form dormant buds wrapped in the dried leaves. The plant comes from southern Mexico and occurs in various communities, even very dry ones. It forms colonies by producing stolons bearing new rosettes which easily take root

322

323

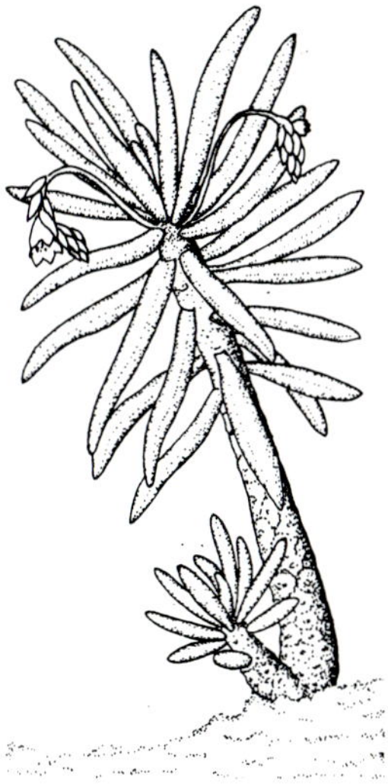

Pachyphytum viride later forms bare, leafless trunks and crowns of green leaves.

324

and found new colonies in the summer. The flowers arise from the rosette at this time; they are tiny but conspicuous thanks to the red-brown colour of their petals.

Graptopetalum filiferum (S. Wats.) Whitehead (325) comes from northern Mexico. It forms abundant colonies of dense ground level rosettes. The leaves taper into fine filiform tips and resemble those of *Sempervivum.* The flowers are relatively large, with reddish tips to the petals. It is propagated from detached leaves or from side-shoots. A closely related genus *Tacitus,* only recently described, differs in its large red flowers. Neither plant has as yet made much impact in the garden or indoor cultivation, but should prove of great value.

There is one genus in the family Crassulaceae which is confined to Europe and the adjacent parts of Asia, that is *Sempervivum,* houseleek, also called 'live-forever'. Although there are only about 30 'true' species, the number of existing names is somewhere in the region of two hundred. Most are cultivars formed by hybridization. Houseleeks are small plants with their leaves in rosettes, widespread in the mountains usually in sunny, dry spots. They form groups and colonies by producing offsets. Cultivation is usually on rock gardens or in pots on balconies, on patios, etc. They require maximum sunshine and abundant watering in summer. They can survive winter kept absolutely dry, but sparse watering is recommended if they are under glass. They are quite hardy. All the species are propagated by rooting single, small side rosettes (offsets), growing on short to long stems. *S. soboliferum,* etc. form tiny bud-like offsets in the leaf axils, easily breaking off and taking root. Some hybrid cultivars are more difficult to propagate.

Sempervivum tectorum L. (326) is, together with *S. montanum,* widely distributed over Europe, and there are many varieties differing in colouration and size of leaves. It used to be planted on house roofs, walls, etc.

325

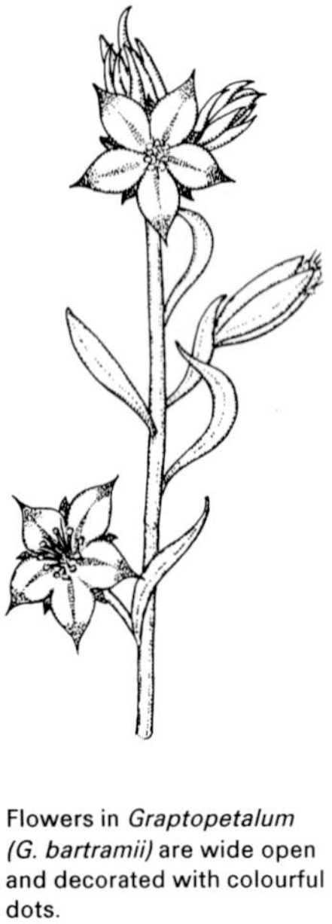

Flowers in *Graptopetalum (G. bartramii)* are wide open and decorated with colourful dots.

326

Sempervivum arachnoideum L. (327), the cobweb houseleek, comes from the Alps, Apennines, Pyrenees and Carpathians. It forms dense cushions of small rosettes, each 3 cm across at the most, the young leaves wrapped in web-like hairs.

Sedum is the largest genus of the family Crassulaceae. Its distribution is not geographically restricted, but it occurs predominatly in the northern hemisphere and also rarely in South America and Africa. *Sedum* includes plants which have creeping, climbing, rosetted or tree-like forms. They may be annual, biennial or perennial. They are not all plants of arid areas, but most of them live in semi-dry to humid locations supplied with water for at least some part of the year. Thanks to their succulence they can survive the less favourable seasons. Many of the representatives of this genus can be encountered in Europe. Those species which are cultivated include plants of cushion-like form, and some later reach a tree-like appearance. The pendent forms grown in hanging pots or baskets are very attractive. In view of the wide range of conditions under which they live in the wild, it is obvious that cultivation is simple, and in many cases these succulents become excellent indoor plants. *S. album, S. acre, S. telephium, S. sieboldii* and many dozens of others, mostly European and Asiatic species, can be grown outdoors.

327

328

Distribution of the genus *Sempervivum*

Sedum nussbaumerianum Bitter (328) comes from southeastern Mexico. It is a plant of sunny spots on limestone rocks in semi-dry woods. It forms small shrubs, up to 25 cm tall, but the stems later become prostrate, spreading out from the original plant. They will take root and increase the size of the plant. New individuals can arise from detached or fallen leaves which will root readily. *S. adolphi* and *S. nutans* are similar species. All thrive in cultivation and can be recommended for window-sills. In summer, they can be planted outside in gardens, in pots on porches, balconies, etc.

329

Sedum rubrotinctum R. T. Clausen (329) reaches 15—25 cm in height. If the erect shoots are regularly snipped off, it forms pads of bright green stems, with tops turning red in the sun. *S. rubrotinctum* comes from higher altitudes of central Mexico where it occurs in coniferous forests, on stony ground, mainly in sunny spots. *S. guadelupense* with green leaves and yellow flowers is a close relative. Some species extend to the north of North America, and they are hardy in the temperate climatic zone, where they survive temperatures below zero. A case in point is

Sedum acre L. (330), undoubtedly one of the best-known species, frequently encountered in more arid regions of Europe and Asia. It seeks rocky, sunny sites on neutral mineral rocks. It produces abundant yellow flowers, in April and May. It is similar to the Mexican species *S. moranense,* which does not tolerate low temperatures.

Sedum hintonii R. T. Clausen (331) is a Mexican species from higher altitudes of northern Mexico. It requires a place in the sun all the year (warmth above 14° C) and careful watering. As with many other succulents, it follows the rule that tougher conditons — all available sunshine, little feeding and sparse watering will result in a more compact growth, better developed white hairs on the epidermis and a more abundant flowering. *S. hintonii*

330

331

flowers in early spring with white blooms. Many other species of *Sedum* are low, prostrate shrubs which can be used individually in interiors, or to complement cactus or succulent collections; if planted on a greenhouse bench, they soon cover large areas.

Sedum allantoides Rose (332) is a representative of the Mexican species. It comes from southern Mexico, from mountain areas. It has a grey-white frosting on the round-tipped, open leaves. It should be cultivated in hard conditions, with all available sunlight and fresh air, and with sparse watering. This will help promote a compact growth and the greatest amount of frosty leaf colouring. It flowers early in summer. A mix of peat, loam and sand is best with a minimal amount of feeding. The winter months can be spent in an absolutely dry environment, providing the temperatures averages 10° C. Like most other species and particularly those from Mexico, it is easily propagated from individual leaves, but with this species propagation is even easier from stem cuttings, often creating aerial roots which root out if the stem is prostrate.

332

333

Sedum album L. (333) is another European succulent plant. It is widespread all over the continent occurring in several varieties and extending to north Africa and to the Far East. Like *S. acre,* it can be grown on rock gardens or in pots standing outdoors for most of the year. It is not particular about soil, tolerates substantial and regular watering, but it does require full sunshine. The white flowers (as the name implies) open in summer and forms with red-tinted blooms also exist.

Sedum morganianum E. Walth. (334), one of the pendent species of *Sedum,* is often used in Mexico as a decorative plant on porches, patios, staircases and gardens. In temperate climates it grows well under glass and makes an excellent hanging basket plant. In summer, the plant needs abundant and regular watering, in winter this can be considerably reduced and temperatures of around 10° C are best though the plant will stand brief spells to 5° C without damage. A close relative, *S. burrito,* has blunt-tipped, smaller leaves of the same grey-blue shade. Both species come from the tropical regions of southern Mexico. They are easily propagated from cuttings or detached leaves.

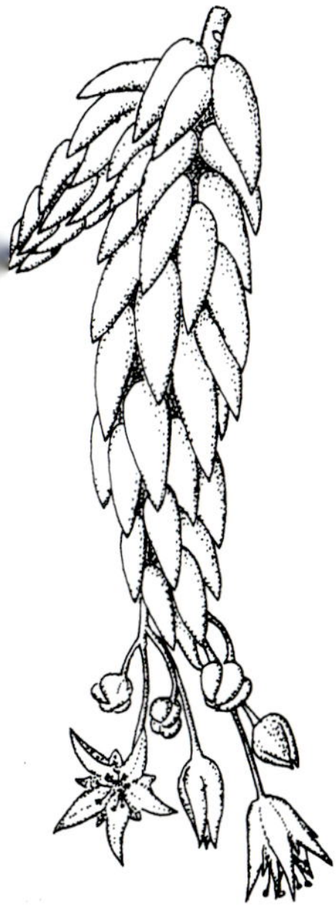

Flowers of *Sedum morganianum* always grow on ends of hanging shoots.

334

335

Sedum frutescens Rose (335). Together with the similar *S. oxypetalum,* it is one of the little appreciated, beautiful succulents, which will certainly soon be introduced into collections. In the wild in central Mexican states, they grow on relatively dry old lava flows, together with other xerophytes. They form shrubs up to 1 m tall with strongly succulent stems covered with an easily removable papery outer skin. The thin tubular leaves are bright green and grow only in favourable humid months, in the same period as the flowers. Throughout the larger part of the -year, the branches are leafless and the crowns look like aged miniature trees marked by tough weather. They are easy to grow. They need a well-drained, moderately rich compost, abundant summer watering and a dry, cool place in winter. Older specimens in particular form beautiful succulent bonsai. Propagation is from cuttings. The leaves are too short-lived to be suitable for propagation.

Sedum pachyphyllum Rose (336) is a prostrate plant, erect and shrub-like in the juvenile state, found among the stones of the famous monumental ruins of the Indian pyramids in Monte Albán (southern Mexico). It resembles *S. allantoides* and differs only in its more robust growth and grey-green leaves with a metallic sheen in the sun. It is excellent for cultivation on staircases, porches, and in summer in gardens.

336

Sedum nevii A. Gray (337) is a native of the states of Illinois and Alabama. In Europe, it flourishes on rock gardens if a porous soil is available. Where sharp changes in temperature after frosty nights are a feature of the climate, winter protection is advisable. *S. ternatum,* a similar species, also comes from North America. It forms rosette-like leaves on sterile, non-flowering stems. Both species produce whitish flowers in summer.

Kalanchoe is an important genus also belonging to the family Crassulaceae. It has over 100 species distributed over the Old World, Africa, and exceptionally extending to India, Indonesia or even to eastern Asia. Some are only slightly succulent and live as clambering lianes in Asian forests. Other species form robust, succulent miniature trees, as does *K. beharensis,* but there are also inconspicuous dwarf species like *K. rhombopilosa.* They are predominantly plants flowering in the short autumn or winter days. They are easily propagated from stem cuttings or detached leaves which can be rooted with little difficulty. Many species form adventitious buds at the tips of leaves often equipped with aerial roots by the time they are shed and quickly giving rise to new specimens.

337

338

339

340

Kalanchoe tomentosa Baker (338) comes from the central mountain ranges of Malagasy, where its low clump-forming shrubs are found on stony and rocky slopes and cliffs. The margins of the oval leaves become marked with dark brown spots in the sun, and are covered with fine felted hairs of whitish colour. *K. tomentosa* is easily propagated from stem cuttings or detached leaves. Shaping the shrubs by careful pruning may later be necessary.

Kalanchoe rhombopilosa Mann et Bout. (339). Many species of this genus come from the island of Malagasy, and *K. rhombopilosa* is one of them. It is a small, shrub-like species with triangular leaves having blunt, toothed and sometimes slightly wavy tips. Their surface is covered with fine grey-white hairs and is marked with brown-red stains which are not always present. Cultivation is as simple as for most species. Warmth and sufficient watering are required in summer, temperatures of about 10° C and sparse watering in winter.

Kalanchoe beharensis Drake del Castillo (340) is a robust species, reaching up to 2 m in the wild. The large and broad leaves are covered with fine felt-like hairs. Forms with hairless leaves also exist. This species comes from Malagasy and its cultivation is easy.

Kalanchoe tubiflora (Harvey) R. Hamet (341) is one of the best-known and most widespread succulents among growers of indoor plants. Tiny buds at the ends of the tubular leaves fall off and develop into new plants. *K. tubiflora* is also indigenous to Malagasy, to the arid southwestern areas, where it frequents the undergrowth of dry forests. It is of importance for collectors of the more delicate species of the family Crassulaceae as it can be used as grafting stock.

341

34

Chapter 11 SUCCULENT SPURGES

The euphorbia family (Euphorbiaceae) includes many tropical and sub-tropical woody plants: some well known house plants including croton and poinsettia; the economically important *Hevea,* from which rubber is obtained, and *Ricinus,* the castor oil plant, and also many temperate species including familiar weeds. There are in all over 5,000 species with a world-wide distribution and about 2,000 of these are succulents occurring in both the New and Old Worlds. These have succulent stems and often bear leaves only in the growing season. The characteristic flower heads are made up of a single female flower reduced to just ovary and stigma, surrounded by a number of separate male flowers, each of just one stamen. There are no petals but the whole inflorescence is completed by a protective ring of joined bracteoles, often bearing glands. The whole is known as a cyathium. In some species, most familiarly the non-succulent poinsettia *(E. pulcherrima),* there are outer bracts which can be large and colourful. The capsules open explosively when ripe, the three segments opening to fling their single seed far from the parent plant. All members of the Euphorbiaceae contain a white sap which is poisonous and must not come into contact with the eye or any other soft, mucous membrane. In strong sunlight it should even be washed from hands and arms.

There are about 500 succulent species in the genus *Euphorbia* itself. Most come from Africa where they are concentrated in the south, though a few grow as far north as Morocco and the Canaries. They can be tree-like with a candelabra-like top *(E. candelabrum);* shrubby *(E. tirucallii);* columnar like the cerei of America *(E. virosa);* shortly columnar or rounded *(E. obesa)* or spreading and more or less prostrate *(E. caput-medusae).* Recently two miniature species which live partly beneath ground level, *E. turbiniformis* and *E. piscidermis,* were found in east Africa to provide collectors with two more desirable plants.

Euphorbia horrida Boiss. (342) is a shrubby species from South Africa, growing in sunny places. It reaches to 1 m in height and forms many side-shoots developing into groups of thorny branches. It is a dioecious species so plants of both sexes are necessary to obtain seeds. *E. horrida* is the host of an interesting and very decorative parasite — *Viscum minimum.* It is one of the smallest mistletoe species, growing only several millimetres long and producing bright red, spherical fruits.

343

Euphorbia horrida var. **striata** W. D. S. (343) is a more slender variety of the species. It has a grey-white, frosted epidermis and shorter thorns. The thorns originate from dried-off flower shoots which have turned woody. Both *E. horrida* and var. *striata* require a porous potting mix and dry conditions in winter. Watering during a cool period is dangerous, as is persistently moist compost. Propagation is both from seed and cuttings of side-shoots.

Euphorbia ledienii Bgr. (344) represents a group of fast-growing shrubby plants, soon reaching large dimensions, up to 2 m in height. In the wild, they form almost inaccessible thickets of thorny branches. It is an undemanding species, fast growing in cultivation and soon producing many yellow flowers. The larger related and often tree-like species *E. trigona, E. lactea* and *E. antiquorum* are distributed over a large part of Africa extending to India. They are frequently used as park and garden plants in tropical and subtropical regions.

344

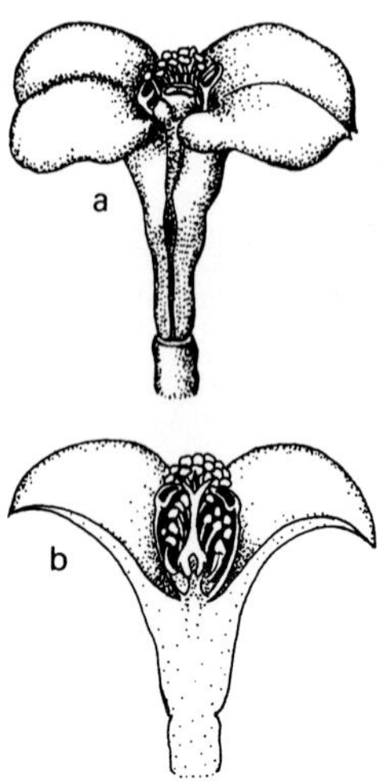

A cyathium in *Euphorbia* has a reduced perianth, replaced by bracts;
a — overall view;
b — cross-section.

345

346

Euphorbia bupleurifolia Jacq. (345) comes from South Africa. It is an unusual species with an ovoid to shortly columnar stem, covered with scars left by deciduous leaves, which grow briefly in the favourable humid season. In the wild, the plant bodies, unprotected by spines, live almost buried in the ground, with only the upper parts and leaves sticking out. The flowers grow on long stalks. In cultivation, the plants bear flowers in a later stage. Propagation is only from seed, because they rarely form side-shoots. They do well in semi-shade. They need regular watering when it is warm, but should be absolutely dry in winter.

Euphorbia caput-medusae L. (346). The species related to *E. caput-medusae* are characterized by their remarkable shape. The body is composed of the main thick stem rising from a thickened, turnip-shaped taproot, anchored deep underground. This stem is hidden by a crown of thin lateral branches, up to 40 cm long and forming a near rosette radiating out like a multi-armed octopus. Only these tuberculate lateral branches can be seen in the picture. The flowers grow on their ends. Another species with a similar 'head' is seen in *E. marlothiana* (348).

Euphorbia tuberculata Jacq. (347). Like the above mentioned 'head' euphorbias, this species also comes from the Cape Province in South Africa. It grows on stony and rocky ground, exposed to the scorching sun.

Euphorbia marlothiana N. E. Br. (348). In the wild, the lateral branches trail on the ground, sometimes taking root. Where this happens, the stem begins to thicken and gradually develops a central boss reminding one of Medusa's head as does *E. caput-medusae.* These species are rewarding to grow as are the related *E. bolusii, E. bergerii* and *E. esculenta.* They tolerate abundant summer watering and a winter place in the light at temperatures above 10° C. They are easily propagated from cuttings taken from lateral branches. The 'head' shape takes a long time to form.

Euphorbia grandicornis Goebel (349) forms shrubs reaching several metres in height. The branches have a most distinctive shape being angled and winged, their edges being armed with thick spines. They grow in pairs and are modified stipules. *E. grandicornis* is easy to cultivate. It requires a warm and sunny place and winter temperatures above 12° C. Propagation from cuttings is easy.

347

348

349

350

351

Euphorbia poissoni Pax. (350) is a shrub up to 1.5 m tall with thick fleshy stems, topped with several elongated, ovate, blunt-tipped and toothed leaves. The surface of the stems is covered with a grey-white, waxy frosting. The species comes from dry mountains in west Africa. It needs a higher temperature than many species and sudden drops below 10° C can be harmful.

Euphorbia unispina N. E. Br. (351) is also native to west Africa. It forms large irregular shrubs, up to 3 m tall. The strongly succulent branches bear leaves only on their tips. The flowers emerge in the humid season, and are borne in clusters. The rest period follows after the leaves have been shed; the plant should then be kept absolutely dry. The species is suitable for greenhouse culture.

Euphorbia obesa Hook. (352) hardly reaches 8—10 cm in diameter. Its form is spherical when young and in the wild, while in cultivation it becomes slightly elongated with age. In the juvenile state, its regularly shaped, green coloured and dark striped stem resembles *Astrophytum asterias.* It is another example of the convergent development of plants in the same climatic conditions. The reduction of the stem in these two unrelated species has progressed in the same way, despite *E. obesa* coming from South Africa and *Astrophytum* from Mexico. *E. obesa* is dioecious. It prefers a sandy, well-drained compost, and tolerates liberal watering in summer, but requires an absolutely dry place in the winter rest period. Propagation is possible only from seed, because it rarely produces offsets and then only when damaged.

352

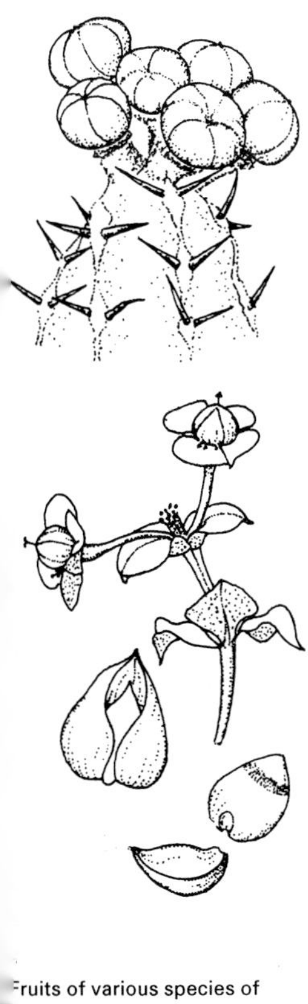

Fruits of various species of *Euphorbia.*

Euphorbia schoenlandii Pax (353) is native to South Africa. It is a robust shrub exceeding 1 m in height. The branches are covered with thick woody spines. These are of different origin to those in *E. unispina* and *E. grandicornis* which originate from stipules, or from *E. horrida* where they are the woody remains of flower stalks. In *E. schoenlandii,* they are lateral vegetative shoots, having leaves in the juvenile state, and shedding them in the dry period, while the stalks remain in the form of 5 cm long and several millimetre thick thorns. The flowers grow on independent, weak and short flower shoots and fall off subsequently. *E. fasciculata* is a similar and related species. In the wild, both plants live in dry semi-desert areas of South Africa, in full sunshine. In cultivation, they require a sunny, warm place and feeding during the summer growing season. In winter, they do best in a completely dry place. They can be propagated from seed or by cutting down older specimens; the base offsets and new shoots will root.

Euphorbia milii Desm. (354) is a well-known shrubby euphorbia from Malagasy. The shrubby species have thinner branches with spines and the thin non-succulent leaves are shed early in winter. In cultivation, *E. milii* is known as *E. splendens* or under a popular name 'crown of thorns' or 'Christ's crown'. Older authors (mainly Ursch and Leandri) described many 'spe-

353

354

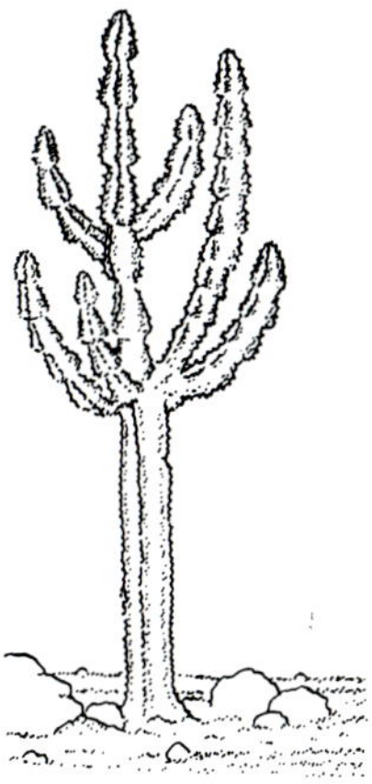

Tree-like euphorbias are a characteristic dominant feature of the African countryside.

355

cies' from the central range of mountains of Malagasy, which, although certainly interesting, are hardly more than forms with differently shaped and coloured bracts. Cultivation is very easy, and the species are often found on amateur-growers' window-sills. An acid compost is recommended, and the plants tolerate liberal watering with additional feed when in leaf. Watering is also necessary in winter if the plants are kept in a warm place.

Euphorbia lophogona Lam. (355) also comes from Malagasy. It forms shrubs up to 0.5 m tall. The branches which have fine spines thicken and become club-shaped towards the crown, and the leaves are larger than in the previous species. It flowers in summer, requiring liberal watering and nourishment. As with *E. milii* the shrubs have to be pruned when they get old, and the prunings will readily put out roots even in water.

Euphorbia piscidermis M. G. Gilbert (356) has a spherical body protected from water losses and overheating by scaly outgrowths. The leaves are absent and the flower shoots emerge only during the short time favourable for growth. *E. piscidermis* and the similar *E. turbiniformis* come from eastern Ethiopia, an area where many miniature succulents have been discovered. Both species are very desirable.

Euphorbia decaryi A. Guill (357) is an interesting, small-sized species from Malagasy with strongly developed tuberous roots and terrestrial branches arranged in clumps. The leaves are tough and leathery, wavy at the edges. Related species, also native to Malagasy, are *E. cylindrifolia, E. françoisii,* and the broad-leaved *E. primulaefolia.* All have thick turnip-like underground tubers and are uncommon in cultivation.

The genus *Jatropha* comes from America and Africa. It covers a number of extremely attractive plants requiring liberal watering and nourishment in summer, and a rest period in winter.

Jatropha berlandieri Torr. (358). Like the previous species, it forms tall stems reaching approximately 0.5 m and growing from a tuberous base. The flowers appear in summer and they are red and highly decorative. It comes from Central America.

356

357

358

359

Chapter 12 THE LILY FAMILY

The lilies (Liliaceae) grow widely all over the world, but their succulent members are found only in the Old World, mainly in Africa. They are leaf succulents, with the leaves arranged in spirals on the stems. The succulent lilies reach large dimensions only exceptionally, as do *Aloe dichotoma* or *A. bainesii.* Their woody trunks grow up to ten metres, and they are the tallest monocotyledonous succulent plants on earth. The growth forms of *Aloe* are very varied; there are species with terrestrial rosettes, but other species have rosettes growing on tall woody trunks while there are also shrubby or miniature species with rosettes of several tiny leaves hidden in the sand of semi-deserts and hardly emerging above the ground. The six-petalled flowers in all the Liliaceae are arranged in two whorls of 3. The seeds in three-lobed (triloculate) capsules are flat, and in some species, they are the only means of propagation in cultivation. The shrubby species and those producing offsets can be propagated vegetatively. Three genera are particularly important for cultivation: *Aloe, Haworthia* and *Gasteria.*

The genus *Aloe* comprises over 270 succulent species, found mainly in south Africa and Malagasy.

Aloe variegata L. (359) comes from South Africa. It forms clumps and colonies by underground offsets and suckers and in cultivation it is easily propagated by separating them from the parent plant. It flourishes on window-sills, and outdoors in summer.

Aloe concinna Bak. (360) comes from the island of Zanzibar in the Indian Ocean. The loose rosettes of white-spotted leaves form numerous clumps.

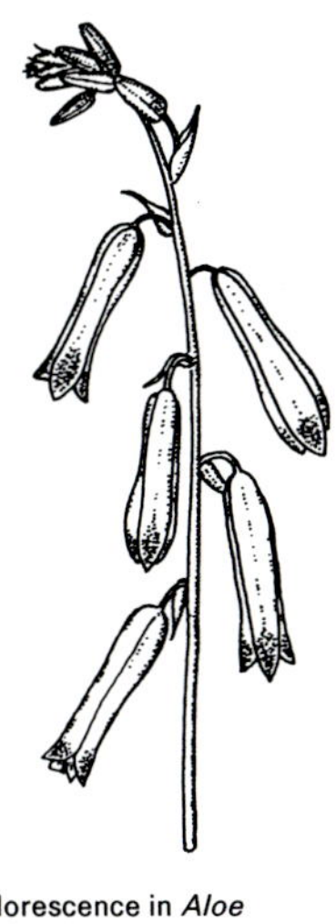

Inflorescence in *Aloe variegata.*

360

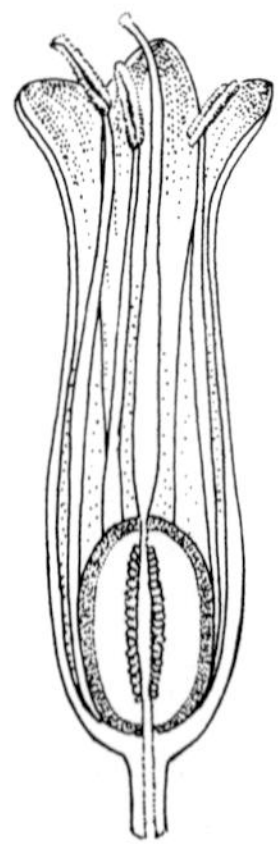

Cross-section of flower in *Aloe variegata.*

361

Aloe rauhi Reyn. (361) is a 10—15 cm high species from Malagasy with light-spotted leaves. The flowers are bright red. In cultivation, it requires a warm place and careful watering.

Aloe millotii Reyn. (362) is a small, elegant aloe from South Africa, frequently found in shrubby undergrowth. It is easy to grow. The pale red flowers appear in summer.

Many species of aloe are used as a source of food; not only those in their native lands, but also the naturalized ones. For instance *A. vera,* coming from the Cape Verde Islands, is nowadays distributed over large areas of Mexico. Its flowers are eaten there. The same applies to the wild species in South Africa, where the flowers are cooked with meat. They are more popular for their texture than for their flavour.

Aloe dewinteri Giess. (363) is a green shrub, easily propagated and in cultivation is very vigorous. The flowers are red and appear in summer.

Aloe melanacantha Bgr. (364) comes from the Cape Province in South Africa. The leaf rosettes, about 30 cm across, are at first terrestrial, later forming a woody trunk, up to 50 cm tall. The leaves, covered with thick thorny outgrowths, are very decorative. The black-tipped spines are soft and do not prick. The flowers are also red, but the species does not flower freely in cultivation. This decorative but rare plant is much in demand. Propagation is normally from seed, only rarely from offsets.

Aloe plicatilis (L.) Mill. (365) is a representative of the larger species and can reach about 2—3 m. It usually looks like a shrub, more rarely forms small trees. The woody branches bear blue-green leaves, arranged in two overlapping ranks. The flowers are bright red. It is a beautiful species but has proved unfortunately difficult to propagate, which prevents it from being seen as often as it should in collections. The woody branches root slowly and the seeds are only rarely available.

Aloe spectabilis Reyn. (366) also reaches almost tree-like dimensions, growing up to about 3 m. The attractive blue-green leaves form a massive rosette, and are up to 1 m long. *A. spectabilis* is a plant suitable only for larger greenhouses. Like the preceding species, it comes from South Africa. *A. polyphylla* from southeastern Africa is another extremely beautiful and decorative species. It forms terrestrial rosettes of spirally arranged, strongly succulent leaves up to 1 m across. It does not produce offsets, and propagation is possible only from seed. It is slow-growing in cultivation, requiring temperatures which never drop below 10° C, and careful watering.

The African genus *Gasteria* with approximately 80 species is of great importance for growers. The plants mostly have two ranked, strong, succulent leaves, covered with whitish stripes, dots or warts. The flowers are inconspicuous, but they are freely borne and they are characterized by the flask-shaped, swollen perianth (*gaster* = belly).

362

363

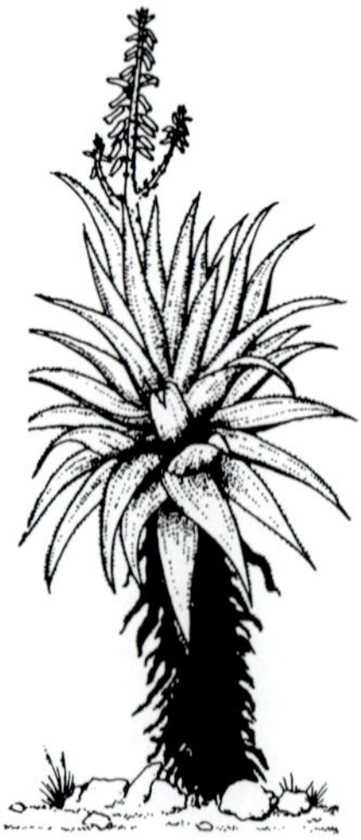

A leaf rosette is the most frequent growth formation in *Aloe;* a woody trunk often appearing. Some species have terrestrial rosettes.

364

365

Distribution of the genus *Aloe*

366

Gasteria maculata (Thunbg.) Haw. (367) comes from the Cape Province, and it is one of the most tough species. It is widely used not only in specialists' collections, but also in the home, where it tolerates lack of sunshine and irregular care. Only permanent dampness and a temperature dropping below 10° C can be harmful. Gasterias are easily hybridized among themselves and with other species (*Aloe, Haworthia,* etc.). Some hybrids are highly decorative.

367

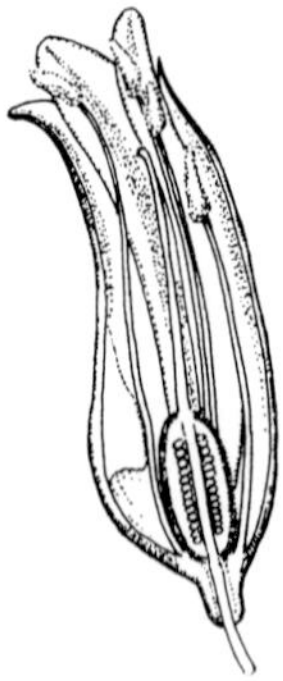

Cross-section of a *Gasteria* flower with the characteristic 'belly-like' tumescence of the lower part of inflorescence.

368

Gasteria armstrongii Schoenl. (368) has leaves arranged in two rows. It is a rare, slow-growing, relatively difficult species, suitable for specialist collections. It requires careful, sparing watering and a dry place in winter.

Gasteria prolifera Lem. (369). The white-spotted leaves form numerous clumps each arranged in two rows and fast growing. *G. prolifera* is one of the easiest species to grow, and its propagation is simple. It is one of the few species where propagation is possible from leaf cuttings, taking root in warm and slightly humid conditions. Because they are undemanding and tol-

369

370

371

372

erant to shading, all the gasterias, particularly *G. verrucosa, G. marmorata, G. nigricans,* and *G. maculata* are species suitable for growing in the home.

Bulbine latifolia Roem. et Schult. (370). *Bulbine* is a relatively small genus from South Africa. Its members have wide open, small, yellowish flowers. Their cultivation is the same as for *Aloe* and *Haworthia. B. mesembryanthemoides* is an interesting miniature species, retracting underground in the hot, dry summer season.

Next to the genera *Aloe* and *Gasteria,* the genus *Haworthia* contains the most varied species in the family Liliaceae. Its about 150 species all grow only in South and southwest Africa. The clump-forming, terrestrial leaf rosette is the only growth form of the genus. The smallest species live buried in the sand of semi-deserts, with only the upper, blunt-tipped parts of leaves emerging above the ground. They are easily cultivated and propagated by detaching the side rosettes. The rarer species are propagated only from seeds.

Haworthia maughanii v. Poelln. (371) is a miniature species from South Africa. It grows virtually buried in the sand. The blunt, flat ends of leaves lie at ground level. They contain transparent water-holding tissue (parenchyma) which lets the sun rays penetrate inside the leaves. They serve as 'windows in the sand'. They are of great importance, because leaves can continue to obtain nourishment even when they are hidden from strong sun rays, yet are not exposed to the dry air and high temperatures. In this way, the succulents with a reduced green surface can continue to use the sunshine through the cells with normal chlorophyll without being damaged. A close and ecologically equally adapted species is

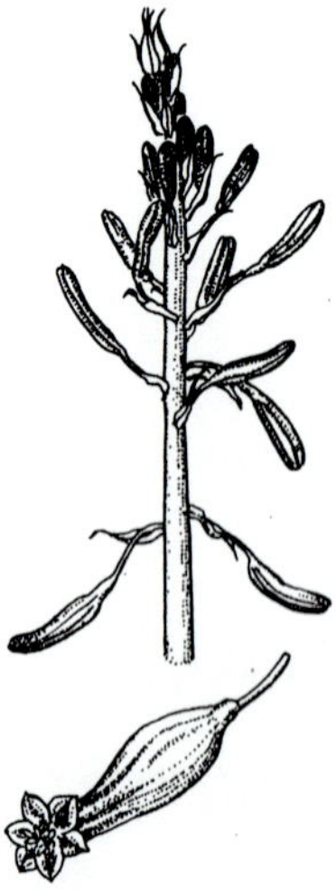

Inflorescence and flower in *Gasteria maculata.*

373

Haworthia truncata Schoenl. (372), having leaves arranged in two rows. It also comes from the sandy semi-deserts of the Cape Province. Both species are in demand, but they are difficult to cultivate. They require careful watering, a porous sandy compost, and a dry spot in winter. Propagation is possible virtually only from seed.

Haworthia cymbiformis (Haw.) Duv. (373) is a small species with transparent, pale green leaves. It is easily propagated not only in greenhouses, but also on window-sills, and in what would be unfavourable conditions for most succulents. It tolerates northeastern or northwestern exposures, where it receives only weak sunlight. It is easily propagated vegetatively, by offsets. It will tolerate abundant, even irregular watering in summer and autumn.

Haworthia fasciata (Willd.) Haw. (374). The small, 5—7 cm rosettes of horizontally striped leaves taper into a soft tip. As with the other species, the flowers are inconspicuous and whitish. This species includes many varieties and forms. Cultivation is easy, as for in *H. cymbiformis.*

Haworthia setata Haw. (375). The leaf rosettes barely reach 5—10 cm across. The leaves have soft, white, transparent hairs on the edges. Several varieties have been described, differing in the size of rosettes and density and length of the leaf hairs. *H. setata* comes from the Cape Province, from dry coastal regions. It requires a very well drained mix in cultivation, careful watering and winter temperatures above 12° C. It is a difficult, prized species, not easy to propagate and rarely encountered in collections.

H. tenera and *H. bolusii* are similar rare species with shorter hairs. In the hottest days of summer they do better in partial shade.

Haworthia attenuata Haw. (376) is perhaps the most common species of haworthia and is very easily grown. It is propagated from detached side-rosettes and suckers and grows fast, forming groups of spotted rosettes marked with small whitish warts. As in *H. fasciata* and *H. tortuosa,* the leaves are stiff, and the plants will survive a long time without watering. However, in the growing season, in summer and autumn, liberal amounts of water should be given.

Haworthia tortuosa Haw. (377) is another widely distributed species with many forms and varieties. Its origin is unknown and it is possible that it is a hybrid, because the individual species of the genus *Haworthia* hybridize very readily, even with related genera. The rosettes of spirally arranged leaves are in *H. tortuosa* green and broadly triangular. Cultivation is easy, the plant tolerating semi-shade and full shade.

The members of the genera *Poellnitzia, Apicra* and *Astroloba* are closely related to and sometimes included in *Haworthia.* They are similar in appearance, differing in the structure of the flowers. Their cultivation is also roughly the same.

Inflorescence in *Haworthia* (*H. tesselata*).

374

Fruits of *Haworthia* species.

375

Bowiea volubilis Harv. et Hook. (378) is a so-called 'root-succulent', developing succulent tissues in its underground bulbs. *B. volubilis* grows in scrubby areas of South Africa, where it is widely distributed, a closely related species extending north to Kilimanjaro. Its bulbs are up to 20 cm across and are usually buried beneath the soil surface. In cultivation, where it is normally planted on the top of the compost the bulb will turn green. In the humid season it grows thin climbing and twining leaves and stems with inconspicuous greenish flowers. Cultivation is similar to that of the easier South African succulents, in winter needing an absolutely dry environment and temperatures around 8—10° C. It can be propagated from single bulb scales treated as cuttings.

376

377

378

Distribution of the genus *Haworthia*

37

Chapter 13 THE LIVING PEBBLES

The Aizoaceae embrace various types of plants, particularly many tropical annual herbs and succulent plants, classified in the subfamily Mesembryanthemaceae. They are distributed, with a few exceptions, in South and southwest Africa. They are leaf succulents, forming shrub-like, mat-forming or clustered plants, and occasionally only individual pairs of opposite leaves making a solitary specimen. The flowers unite the group. They are radial, many-petalled with numerous stamens, reminding one of daisies *(Bellis)*. The fruits are mostly dehiscent capsules. Water is an important factor in releasing the seeds. During rain and on dewy nights, the capsules open and the tiny seeds are washed out.

About 2,000 succulent species of the family Aizoaceae are mentioned in literature, but contemporary revisions suggest that the number should be lower.

Growers are most interested in the dwarf species, the so-called living stones, with globular, club-shaped or ovoid leaves, mostly in pairs. In the genus *Argyroderma,* their surface is smooth, mostly grey-white. The bright coloured flowers appear terminally between each pair of leaves and come in shades of yellow, white and red. All the species of *Argyroderma* are native to South Africa. They live in fine alluvial sediments, almost buried in the soil. According to recent monographs, the genus *Argyroderma* includes only about 10 species.

Argyroderma schlechteri Schwant. (379). As with all the other species, this one grows in clumps, but it is slow-growing and it is many years before the pairs of opposite leaves develop into clumps. The dull, grey-white leaf surface is inconspicuous for most of the year, only becoming obvious when the red flowers appear at the end of summer.

Argyroderma carinatum L. Bol. (380) is chiefly solitary, only rarely forming clumps. According to reference books, the flowers of this species are pinkish red, but many authorities have confirmed their considerable variability of colour.

Argyroderma octophyllum (Haw.) Schwant. (381) is one of the most rewarding species, widespread in cultivation mostly under the wrong name of *A. testiculare.* It

380

381

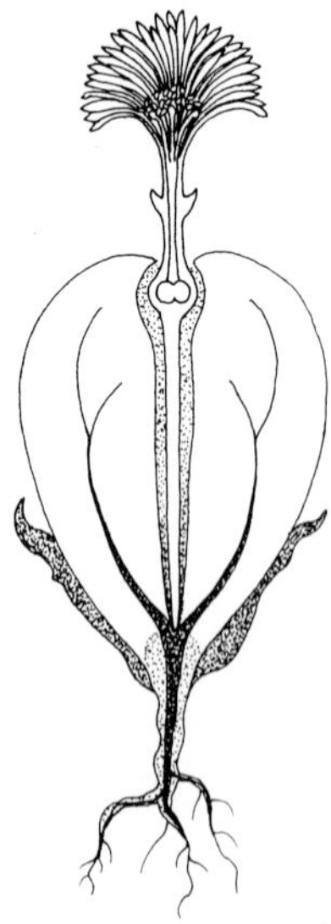

In *Conophytum,* flowers grow through slits in leaf pairs.

has yellow flowers and leaves with a glaucous patina. Cultivation of argyrodermas is not difficult: they thrive in a well-drained, sandy compost in a sunny spot. Watering is the biggest problem. The growing season takes place in late summer and autumn, while spring and high summer are a rest period for the plants. Therefore they do not need watering before August and September. If they are watered before that, it should be only on warm days and then sparingly. Propagation, as with other members of the Mesembryanthemaceae, is possible only from seed, or by dividing the clump-forming plants. The best time to do this is August.

Conophytum is a large genus of some 300 species. The reduced leaves are almost globular, with a tiny slit in the middle of the leaf pair. The flowers emerge early in autumn from this slit. Most of the year, the leaves are covered with the membranous remains of the old, dead leaves, and to a layman, they look like plants which have been dead for a long time. When mature, they form clumps, often groups of many dozens of leaves. There are several species which form elongated stems in age. *Conophytum* comes from South Africa, mostly frequenting stony and rocky hills. The clumps of tiny bodies hide from the sun among stones and in crevices, growing in sedimentary clay or sand.

Conophytum praegratum Tisch. (382) is native to the sandy hills of Namaqualand. It forms numerous clumps of small pairs of leaves approximately 10 mm wide. The dark dots on their surface do not serve as ornaments or mimicry. They are the so-called 'windows' letting through diffused light into the deeper chlorophyll-rich layers of the leaves. As early as last century, Schoenland proved that the leaves of species equipped with these tiny

382

383

'windows' have under the surface a much stronger layer of the green pigment chlorophyll, the so-called assimilating parenchyma, than the species without such 'windows'. They are often seen in the genus *Conophytum.*

Conophytum spirale N. E. Br. (383) differs from the preceding species in having flatter leaves and yellow-white flowers (*C. praegratum* has pink blooms). The flowers arise in early September. The plant comes from the Cape Province, where it grows in sandy alluvial sediments.

Conophytum wettsteinii (Bgr.) N. E. Br. (384) is a small species with leaves about 2.5 cm broad tending to form clumps. In the wild, it is a plant of rocky crevices. In the dry season the green colour of the live leaves is covered with the grey-spotted coating belonging to the epidermis of the previous year's leaves. The flowers are a striking carmine colour.

Conophytum cupreatum Tisch. (385). Even though the name of the species is derived from the copper colouration of the flowers, the plant is also characterized by elongated cylindrical leaves and a long stem which spreads underground, producing sucker growths which help to form wide mats.

Conophytum cordatum Schick et Tisch. (386) comes from the southern Cape Province, from mountain areas. As with the preceding species, the leaves are heart-shaped, with deep notches between them. The yellow flowers emerge in late summer and in autumn. All the above-mentioned species have blossoms which open by day.

Distribution of the genera *Conophytum* and *Lithops*

384

385

Conophytum scitulum N. E. Br. (387) is also native to the southern Cape Province. It forms dense clumps of tiny leaves. The whitish flowers open in the evening and remain open until the morning. They are scented as are all the nocturnal flowered species.

Cultivation of all the *Conophytum* species means adapting the treatment given to fit in with their natural rhythm of growth. Summer and spring are the rest period, when life is sustained by occasional watering, but permanent moisture would kill the plant. The beginning of the growing season is apparent when the old epidermis cracks between the double leaves. At that time, watering is necessary. To interfere with the root system is also risky, and the plants should not be repotted for several years. Propagation from seed requires patience; the plants are slow growers. In the first seasons, the seedlings require watering throughout the year. In the rest period, temperatures should be kept above 10° C.

Many members of the Mesembryanthemaceae are wide-spreading plants forming clumps or mats. They have elongated, trailing stems, often rooting as they spread. The flowers have striking, bright and lustrous colours. Unlike the miniature genera *Conophytum, Argyroderma* and *Lithops,* which are to a certain extent difficult to cultivate, these plants are good growers. Spring and summer are the period of maximum growth, but they will continue growing in autumn and winter if watered. They can be used to cover larger surfaces, in greenhouses and outside in summer. They are easily propagated from terminal cuttings with several pairs of leaves.

386

387

388

Cheiridopsis turbinata L. Bol. (388) comes from South Africa as do most of the species in this genus. They live in the sandy soil of flat, open country. In cultivation the pale yellow flowers are borne in abundance.

Cylindrophyllum calamiforme (L.) Schwant. (389). The whole genus comprises about 5 similar species, native to the Cape Province. They form clumps of thin stems and cylindrical leaves. The flowers are pink and white. Cultivation is easy, requiring watering in summer, and a dry, light place in winter.

Delosperma cooperi (Hook.) L. Bol. (390). *Delosperma* is a genus with many dozens of species also coming from South Africa. They have shrubby to trailing and creeping stems, numerous leaves and white or red flowers. They are easy to grow and some species, for example *D. tradescantioides* and *D. taylori,* can be put outside as hanging basket plants on balconies in summer.

Cerochlamys pachyphylla (L. Bol.) L. Bol. (391). The systematic arrangement of the subfamily Mesembryanthemaceae is based mainly on the structure of the fruit and because it requires microscopic study it is a matter for the expert not the beginner. It may seem that the single species of the genus *Cerochlamys* has a lot in common with the genus *Mesembryanthemum*

389

390

and with the above mentioned genera, but in fact the fruit structure is quite distinct.

C. pachyphylla has rosettes of triangular leaves and red flowers. Its growth season is in summer.

Faucaria bosscheana var. **haagei** (Tisch.) Jacobs. (392). The genus *Faucaria* comprises several dozens of interesting and rewarding species, called 'tiger's jaws', from the toothed edges of the leaves. *F. tigrina* is the best-known species. In age, all the species form small shrubs, producing yellow or pink flowers at the end of summer, and they are easily propagated both from seeds and by separating individual rooted rosettes. During winter they require an absolutely dry and light place at temperatures above 10° C. The distribution of the species is restricted to the Cape Province.

Fenestraria rhopalophylla (Schltr. et Diels) N. E. Br. (393). The name of the genus is derived from the Latin word *fenestre,* that is window. The windows are visible at the ends of the club-shaped leaves like darker spots. They allow the sun's rays to penetrate into the leaves, where they are used for photosynthesis. Unlike the small windows of *Conophytum,* these windows are large and let through more light. The plants live almost completely hidden underground, with only the blunt tips of leaves at ground level. Cultivation of fenestrarias is not difficult; it is important, however, to see that they have a rest period in early spring and early summer. The whitish flowers emerge in autumn.

391

392

393

394

Fenestraria aurantiaca N. E. Br. (394) differs from the preceding species in having yellow flowers. Both species succeed in a sandy, loam-based compost, and require watering from mid-summer to autumn. Propagation is from seeds.

Herreanthus meyeri Schwant. (395) comes from the Cape Province. It is a plant of sandy soil in flat alluvial ground. It has fleshy triangular leaves. The snow-white perfumed flowers open during the night and continue to do so for up to a week. They are highly decorative. When mature, the plant forms small clumps. It is easily cultivated on window-sills in a sandy potting soil. The end of summer and autumn is the main growth season when the plants require maximum watering and warmth. Propagation is only from seed.

395

396

Frithia pulchra N. E. Br. (396) is a small rosette plant with leaves equipped with developed windows. The flowers are purplish-pink. The plant is native to sandy, flat country in the Transvaal. Cultivation is the same as for *Fenestraria.*

Dorotheanthus bellidiformis (Burm.) N. E. Br. (397) comes from South Africa, and it is cultivated all over the world. As a summer plant, it is used in dry and warm sites, quickly produces clumps, and dozens of flowers in high summer. It is propagated annually from seeds which are widely available. There are cultivars of many colours. *Drosanthemum* is a similar but perennial genus. It includes a dozen shrub-forming species, often prostrate and creeping, with long leafy stems, bearing colourful flowers in the summer months. The South African *Carpobrotus* is also common in cultivation. *C. edulis* is the most familiar species: its fruits are picked and eaten like figs, because they do not dry up at maturity but remain fleshy. *Carpobrotus* has been planted along many coasts to strengthen the

397

sandy dunes. Due to its fast growth and formation of fleshy leaves and stems, experiments have also been carried out to use it as cattle fodder. *C. chilensis* is the only member of the succulent Mesembryanthemaceae to be indigenous to the New World. All the species of the genus *Carpobrotus* produce beautiful flowers and they are easily cultivated. They are decorative in large greenhouses, needing space to form mats.

The genus *Gibbaeum* comprises some 20 species, all native to the Cape Province. They are small, clump-forming succulents with ovate to globular, rarely elongated pairs of leaves, smooth on the surface and lacking windows. In some species, for example *G. dispar* and *G. pachypodium,* the leaves are variable in size. The flowers are mostly pink to red, rarely whitish. They appear in autumn or spring, depending on the species and how it is cultivated. The period of growth varies: some species grow in autumn and winter (for example *G. perviride, G. pilosum* and *G. velutinum*), others continue to grow until the spring or summer months (*G. cryptopodium, G. haagei, G. heathii,* etc.).

Gibbaeum dispar N. E. Br. (398). The pairs of unequally large leaves later form small clumps. The purple-red flowers emerge in August before the plant starts into growth. In September and October, the leaf pairs begin to grow. At this time, careful watering, maximum sunlight and warmth are necessary. In

398

399

favourable conditions, the growth will continue into winter, but since it is difficult to provide the plants with sufficient sunshine and warmth, it is better to induce a premature rest period by restricting the watering. If the compost stays moist when the temperature drops, the roots can be damaged. The plants survive the warm summer season in a state of rest. They are easily propagated from seeds. The seedlings grow fast and bear flowers in the second or third years.

Gibbaeum schwantesii (N. E. Br.) Tisch. (399) is one of the rarest species in the wild. It has elongated, mostly prostrate to creeping stems and longish, keel-shaped, bright green leaves. *G. velutinum* is similar but has red flowers, while those of *G. schwantesii* are snow-white. Both species are propagated from stem cuttings which root easily. Autumn is the growth season. No species of the genus *Gibbaeum* tolerates temperatures remaining below 10° C for long, especially if the compost is moist or humidity high.

Lithops is the best-known genus of the succulent Mesembryanthemaceae. It has about 80 species of tiny, usually clump-forming plants. The

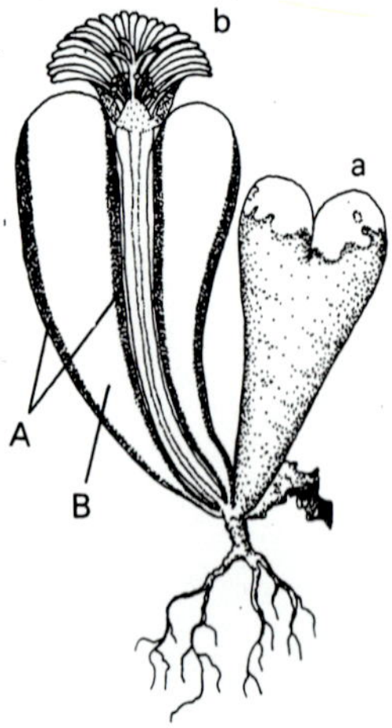

Two leaf pairs in *Lithops*;
a — transparent windows in blunt leaf tips;
b — cross-section
(B — water parenchyma;
A — chlorophyll).

400

401

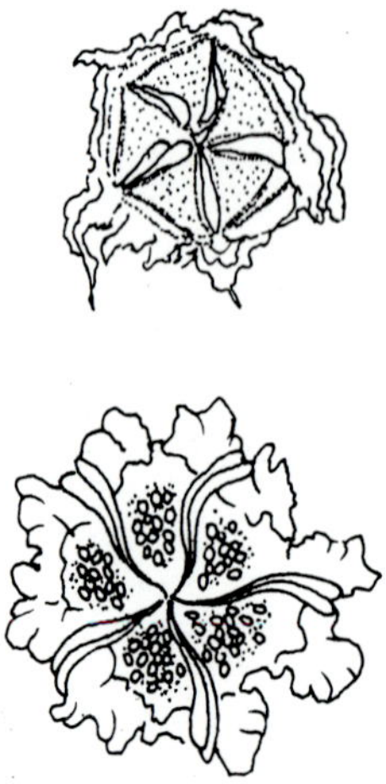

Fruits in *Lithops* open in a humid environment.

402

403

pairs of reduced leaves are formed in a similar way to *Conophytum,* with a slit in the middle from which the flower grows. The pairs grow afresh every year: the new pairs develop on a short stalk and gradually grow between the old pairs, finally equalling or exceeding them in size. The old leaves are then no more than a cover protecting the young leaves from the sun. In *Lithops,* the windows in the leaves are also encountered as tiny dots (as in *L. gracilidelineata* and *L. fulviceps*), or as interesting patterns on the upper side of leaves *(L. aucampiae, L. olivacea).* In the wild the plants occur on flat, alluvial ground, and live almost buried in the soil or in rocky crevices. They are not only perfectly camouflaged by the fine sediments (mimicry), but are also virtually cemented in. All the species live in South Africa. In cultivation they are not difficult requiring full sun and water in summer when the compost is dry. Always err on the side of dryness when in doubt. Over-watered plants grow tall and ugly. They flower at the end of the growing season in late summer to early autumn and thereafter can be dried off for the winter. A winter minimum of 5–7° C is adequate for most species when in this state.

Lithops bella (Dtr.) N. E. Br. (400). The dark spots or larger stains and lines are in fact translucent cells through which the light penetrates into the leaves. In nature, the pattern on the leaves will blend with the surroundings, and in the drought, when the plants bear no flowers, they look like pebbles, and the name 'living pebbles' or 'living stones' is appropriate. Many species are characterized by the pattern on the upper side of leaves.

404

405

Lithops julii (Dtr. et Schwant.) N. E. Br. (401). The surface of the leaf pairs is grey-white, without the dark windows. Even the described varieties, var. *pallida* and var. *reticulata,* lack the pattern of dark spots on the upper side of leaves. Dark brown spots along the slit between the leaves are, however, often visible. The white flowers grow early in autumn.

Lithops fulviceps (N. E. Br.) N. E. Br. (402) comes from southwest Africa, from Namaqualand. The leaf pattern is brown-red, but it develops only in adult plants. In juvenile state, the leaves have only dark dots.

Lithops dorotheae Nel. (403) has a characteristic pattern on the upper side of leaves and yellow flowers. It comes from South Africa, from heavy, clayey, alluvial soil.

Lithops aucampiae L. Bol. (404) is native to the Transvaal. It is a denizen of flat sites around the dried-up seasonal river beds, through which water flows only in times of flood. The flowers of this species are also yellow. *L. aucampiae* is common in collections and easy to grow. It does not differ in cultivation from the other species. The flowers appear from August to October.

Lithops gracilidelineata Dtr. (405). The thin, hair-like, intricate leaf pattern is brown-red. After some years, this *Lithops* forms larger groups and flowers easily (as do the preceding species) in early autumn, bearing yellow blossoms. It is found on broken granite gravel in a matrix of clay, in the deserts of Namaqualand.

406

407

408

Lithops lesliei (N. E. Br.) N. E. Br. (406). The leaves are about 2 cm across, with a striking brown-red pattern. The flowers are yellow. Propagation of *Lithops* in cultivation is easy from seed and larger clumps can be divided. In the first months, the seedlings need diffuse light, never direct sunlight, a nourishing, humus-rich compost, and permanent moisture. Later, when they are pricked off they need good ventilation until they are used to full sunlight. Watering is probably the most important factor: the first rule is that the plants are better with too little watering, than with too much. Watering should follow a thorough drying of the compost, lasting for several days. They are best kept completely dry in winter when most will survive temperatures down to 5° C.

The genus *Pleiospilos* with several dozens of species includes clump-forming to shrubby succulents with conspicuously developed fleshy leaves and large yellow flowers. All come from South Africa. The flowers appear in early autumn. Most species thrive if treated like *Faucaria* or *Gibbaeum.*

Pleiospilos simulans (Marln.) N. E. Br. (407). The leaves are up to 4 cm wide and long, and the flowers are of the same diameter. The blooms have a pleasant fragrance. *P. magnipunctatus* and *P. bolusii* look similar. They are propagated from seeds, and the seedlings reach flower-bearing size within 2—3 years.

Ophthalmophyllum lydiae Jacobs. (408). The name of the genus indicates the appearance of the plant in nature, where the rounded ends of leaves emerge above the sandy soil like fish eyes. The translucent epidermis at the ends of the leaves and the water-storage parenchyma below enables the plant to remain otherwise underground.

South and southwestern Africa are the home of the genus *Titanopsis,*

409

410

with its 8 species of tiny clump-forming plants. The leaf surface i fissured like a piece of rough rock, perfecting its camouflage. Yellov flowers are produced in autumn. The plant requires a porous compos and careful watering, being kept dry in the winter and spring rest period. It is a very rewarding succulent and can be cultivated on window sills in the home.

Titanopsis schwantesii (Dtr.) Schwant. (409) and

Titanopsis calcarea (Marl.) Schwant. (410) are the best-known species, with whitish tubercles a the tips of blunt leaves. They are propagated from seeds or by divisio of larger clumps.

South Africa is also the homeland of the genera *Nananthus* (16 species) and *Rhinephyllum* (14 species). They also have rosettes of opposite leaves, later becoming shrubby. The larger part of the plant lives i the soil, and only the tips of leaves stick out on the surface. The flower are usually yellow (in *Rhinephyllum*), or pink (in *Nananthus*).

Nananthus malherbei L. Bol. (411). The tips of leaves bear whitish tubercles, in the wild camouflaging the plant in limestone gravel. Summer and autumn are the season of growth, when the flowers open. Like *Titanopsis, Rhinephyllun Rabiea* and *Pleiospilos,* it thrives in limestone soil. Sometimes foun listed as *Aloinopsis.*

411

Rhinephyllum broomii L. Bol. (412) is an easily grown succulent. It requires liberal watering in summer. In winter when temperatures are lower the plants should be kept dry. Propagation is from seed or by division of larger clumps, in late spring.

412

413

414

Chapter 14 SUCCULENTS OF OTHER FAMILIES

Succulent plants can be found in many other families but there are too many to be described in one book. Priority is therefore given to those which are best in cultivation because of their relatively smaller size or their undemanding cultural requirements.

In the purslane family (Portulacaceae), there are two such genera: *Anacampseros* and *Ceraria.*

Anacampseros species form miniature plants, sometimes with a strongly developed stem base (caudex). The species come from Africa with the exception of *A. australis* from southwestern Australia.

Anacampseros albissima Marl. (413). The stems are about 4 mm in diameter, and they are densely covered with papery scales which cover the tiny green leaves. As it ages the plant forms clumps. The flowers appear in summer. Spring and summer are their growth season, when the plants need abundant watering and a place in full sun. In winter, they have to be kept dry, and are best at temperatures above 7° C.
Portulacaceae

Anacampseros crinita Dtr. (414). The stems grow to 8 cm in height, and groups and clumps are formed by offsets. The leaves are fleshy and tubular. It is the easiest species in cultivation. A well drained sandy compost is suitable. *A. crinita* is relatively easily propagated by cuttings, whereas most other species, especially those with tuberous roots or stem bases, can be propagated only from seeds.
Portulacaceae

415

Leaves in some *Peperomia* species *(P. dolabriformis)* are longitudinally folded; sunlight enters only through a narrow groove.

416

Ceraria namaquensis (Sond.) Pears. et Stephens (415). The genus *Ceraria* comprises only a few
Portulacaceae
species. All are found in South Africa. *C. namaquensis* is a shrubby plant, up to 2.5 cm tall in the wild, with tiny ovate to tubular leaves. They can be shaped into bonsai. *C. pygmaea* is an interesting stunted species, developing a tuberous base to the stem. The leaves of both species persist only in favourable conditions and fall off when the plant is over-wintered in a cool place. The flowers are inconspicuous. Propagation of both species is possible from cuttings, but it is faster to graft them on the more resistant *Portulacaria afra* which is easily propagated vegetatively. In summer, the plants tolerate abundant watering. They should be kept absolutely dry during winter, at a temperature above 10° C.

In the family Peperomiaceae we encounter succulents of the genus *Peperomia,* found throughout Central and South America. In the wild peperomias usually occur in permanently moist communities, some even as epiphytes in cloud forests. Many are cultivated as house plants, the succulent species, *P. dolabriformis, P. columella, P. galioides,* and *P. nivalis* less so. The succulent types have the water parenchyma in the leaves on the upper side, and some even fold their leaves to form a narrow, translucent window on the upper side.

Peperomia graveolens Rauh et Barthlott (416) comes from the Peruvian mountains, where it
Peperomiaceae
often grows on bare rocky walls in moss cushions. It branches out to

form a shrub and reaches several decimetres in height. The distinct reddish leaves have the translucent tissue on the upper side as can be seen in the picture. In cultivation they require a humus-rich compost, sufficient watering in the warm season, and careful watering in winter, when temperatures should not drop below 14° C.

South Africa is the homeland of the genera *Sarcocaulon* and *Pelargonium,* belonging to the family Geraniaceae. Few people, cultivating species of *Pelargonium* (geraniums) on their window-sills know that there are over 250 species in this varied genus, inhabiting the dry regions of southwestern Africa, and that many of them are stem-succulents.

Sarcocaulon spinosum (Burm.) O. Ktze. (417) forms thickets of spiny stems over 1 m tall. The white flowers grow singly, distinguishing them from the genus *Pelargonium.* In other species, they can be yellowish to pinkish red. Other species, suitable for use as bonsai, are *S. herrei* with yellow flowers, *S. multifidum,* lacking the spines, and *S. rigidum,* resembling *S. spinosum* in the picture. The genus comprises some 15 species.

Geraniaceae

Pelargonium tetragonum (L. F.) l'Her. (418) reaches about 1 m or more. At that size, the main stem turns woody and the lateral square branches are about 1 cm across. While the segmented succulent stems are permanent features, the leaves grow only in the humid season. The species is easily grown and propagated in cultivation.

Geraniaceae

Most species of the genus *Sarcocaulon* and of xerophilous pelargoniums are more difficult to cultivate, the plants coming into leaf in late autumn. In summer, they should be grown in plenty of fresh air and

417

418

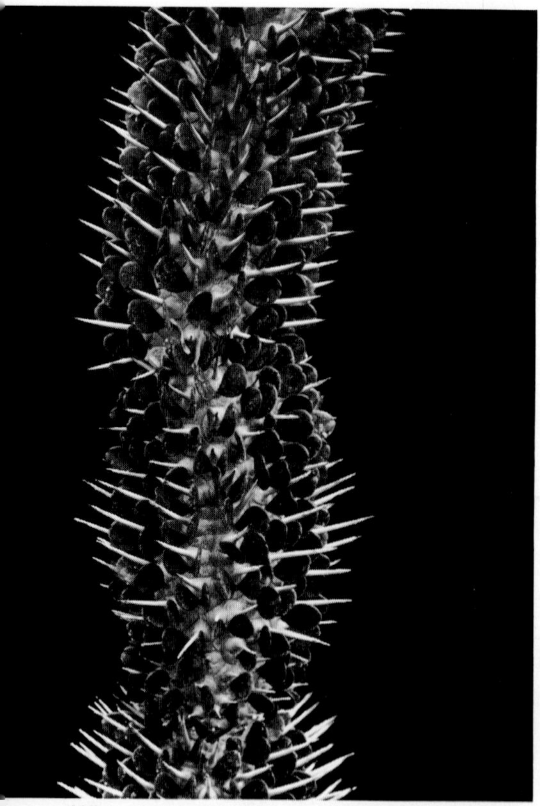

419

420

watered liberally and in winter need good light. They can be propagated from soft wood cuttings taken from branches that have not yet become woody.

The family Didiereaceae comes from Malagasy. Most of its members are large, tree-like plants. Some can be easily shaped and cultivated as bonsai.

Alluaudia montagnacii Rauh (419). Although in nature it is a tree reaching many metres in height, it can be shaped into acceptable dimensions in cultivation. Equally suitable for cultivation are *A. ascendens* and *A. procera.* They can be propagated from branch cuttings.
Didiereaceae

Only a few members of the family Didiereaceae will bear flowers in cultivation. *Alluaudiopsis marnierana* with thin, zig-zag stems is one of them.

Bursera fagaroides Engelm. (420). The family Burseraceae comprises several genera of desert trees, for example *Commiphora* and *Terebinthus* from Africa and Arabia, but of greatest importance for the growers is the genus *Bursera* itself with more than a hundred species, native to tropical and subtropical America. They are mostly forest trees. Some Mexican members
Burseraceae

form exceptionally thickened woody trunks (caudex). The tissue of the trunks is soft and porous, containing channels filled with watery resin. Many species are accordingly exploited for their resins (copal, elemi, tacamahaca).

B. microphylla from the western coast of Mexico and *B. hindsiana* from Baja California make equally interesting stunted trees in cultivation as *B. fagaroides* and *B. odorata* do from central Mexico. They are easily cultivated in ordinary well drained soil. In summer, when they have leaves, they require plenty of water, while in winter they tolerate a dry environment. Propagation is best from seeds. Cuttings make exuberant growth producing large shrubby plants, but the thick caudex is not formed.

East Africa is the home of the genus *Dorstenia* of the family Moraceae, which includes some stem-succulents, mostly with fleshy, bottle-shaped, thickened stems. Their cultivation is more demanding needing a permanent supply of water, and continuous high temperatures. They are not suited for the average cactus greenhouse. They will flourish in indoor glass plant-cases or specially heated greenhouses.

Dorstenia gigas Schweinf. (421) comes from the Socotra Island in the Indian Ocean. The bottle-shaped trunk reaches 1.5 m in height and 0.5 m in diameter.
Moraceae

421

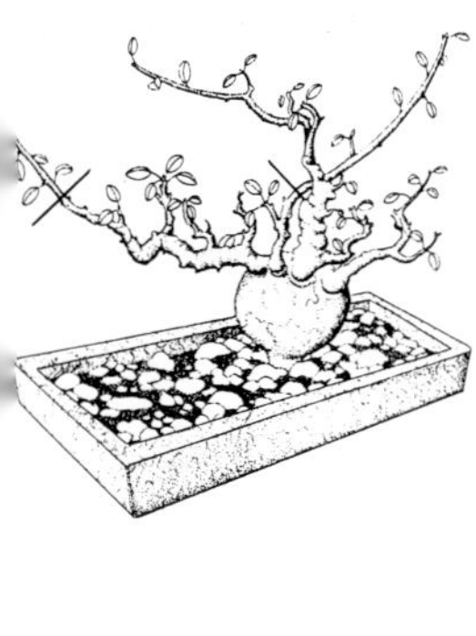

In bonsai shaping, for example in *Bursera microphylla*, long shoots have to be cut off.

Distribution of the genus *Dorstenia.*

422

423

Dorstenia crispa var. **lancifolia** Engelm. (422) is found amongst the undergrowth of dry, deciduous forests in east Africa. The succulent stems reach about 20 cm in height, and often form clumps with age, particularly in the favourable conditions of cultivation. The leaves with their wavy margins are highly decorative. The flat, lobed inflorescences bear small, densely arranged flowers. They open in high summer, and often again in autumn.
Moraceae

Dorstenia foetida Forks. (423) comes from southern Arabia. It is the best-known species. In the wild, the plants are only several centimetres tall. The elongated stems and clumps are met with only in cultivation. For successful cultivation they require a sandy, humus-rich compost and all-year round higher temperatures. In winter, watering should be reduced.
Moraceae

Oxalis carnosa Mol. (424) represents the family Oxalidaceae, in which succulence of stems and roots is encountered in several South American species belonging to the genus *Oxalis. O. carnosa* comes from the dry coastal areas of Peru and Chile, ascending to higher elevations in the Bolivian mountains. The tuberous storage roots and succulent stems enable the plant to survive several months of drought, when the shrubs are leafless. Together with *O. sepalosa,* they can be shaped as bonsai. In cultivation, they grow successfully; spring and summer is the growing season, and in winter,
Oxalidaceae

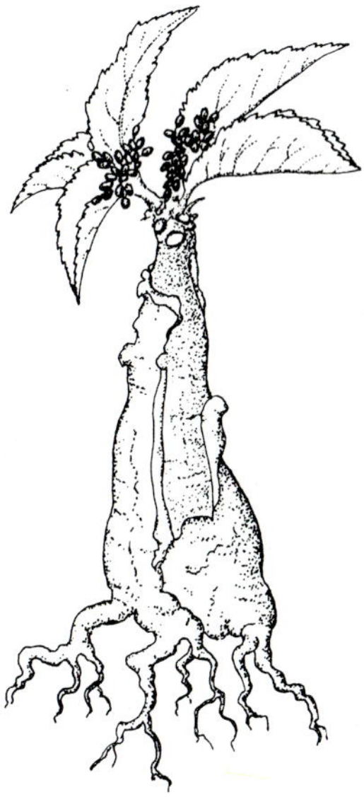

Succulents such as *Cyphostemma* have thickened stems making a natural bonsai shape.

424

they need a completely dry place. Temperatures can drop below 10° C. Propagation is vegetative, lateral branches take root easily.

The original genus *Cissus* (family Vitidaceae) is now divided into two genera. The present-day genus *Cissus* includes climbing and twining succulent lianes with tendrils and segmented bodies, while the shrubby, massive succulents with papery separable bark, are put into the genus *Cyphostemma.* The flowers in both genera are arranged in simple clusters, and the fruits are berries very like those of the vine. *Cissus* and *Cyphostemma* are robust succulents, unsuitable for collections with restricted space; they can be grown only in large greenhouses. They quickly adapt to cultivation, requiring a lot of water in summer, and nutrients for exuberant growth. They spend the rest season (winter) completely dry and leafless, ideally at temperatures of about 10° C.

Cyphostemma juttae Dtr. et Gil. (425). In the wild, it reaches over 3 m in height, and the thick succulent trunks are over 1 m wide at the base. They are covered with light, papery bark, peeling off in rings. *C. bainesii, C. crameriana* and *C. elephanthopus* are similar. The species from Malagasy, *C. laza* with the bottle-shaped trunk, produces thin, clambering branches. In cultivation, it is easily propagated by rooting branches or from seeds.
Vitidaceae

Cissus quadrangula L. (426). The quadrangular climbing stems are segmented. In the wild, it grows like a shrub with clambering branches, holding by its tendrils onto tree tops and bushes. The similar *C. cactiformes* also comes from South and southeastern Africa.
Vitidaceae

425

426

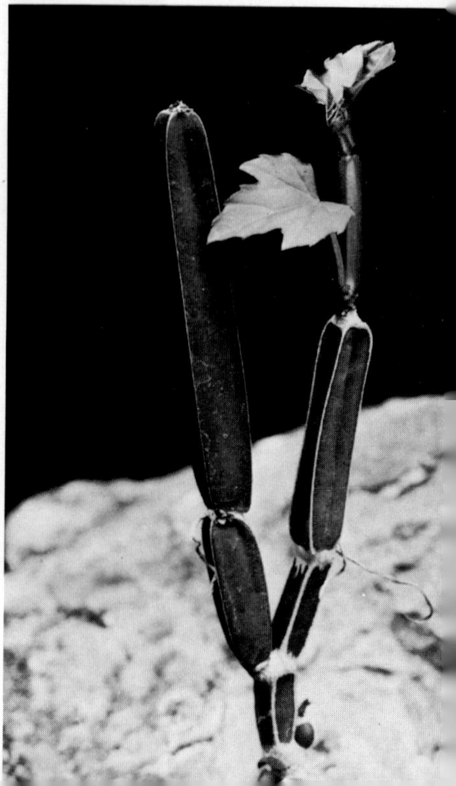

BIBLIOGRAPHY

Backeberg C. (1978): Cactus Lexicon, Blandford Press Ltd., Dorset
Barthlott W. (1978): Cacti, Stanley Thornes Ltd., Cheltenham
Barkhuizen B. P. (1977): Succulents of Southern Africa, Purnell and Sons, Cape Town, S. A.
Bayer M. B. (1977): Haworthia Handbook, Kirstenbosch, S. A.
Bechtel H. (1977): Cactus Identifier, Oak Tree Press, New York
Ginns R. (1975): Cacti and other succulents, David and Charles, Devon
Glass C., Foster R. (1976): Cacti and Succulents for the Amateur, Blandford Press Ltd., Dorset
Innes C. (1977): The complete handbook of cacti and succulents, Ward Lock Ltd.
Jacobsen H. (1974): Lexicon of succulent plants, Blandford Press Ltd., Dorset
Lamb E. B. (1972): Pocket Encyclopaedia of Cacti in Colour including other Succulents, Blandford Press Ltd.
Lamb B. E. (1974): Colourful Cacti and other succulents of the Desert, Blandford Press Ltd.
Noble W. C. (1976): Aloes for Greenhouse and Indoor Cultivation, NCSS
Oudshorn W. (1977): Cacti and Succulents in Colour, Lutterworth Press
Rauh W. (1979): Bromeliads for Home, Garden and Greenhouse, Blandford Press Ltd.
Rauh W. (1979): Die grossartige Welt der Sukkulenten, P. Parey, Berlin—Hamburg
Rauh W. (1979): Kakteen an ihren Standorten, P. Parey, Berlin—Hamburg
Rice P. (1976): Cacti and Succulents for Modern Living, Merchant Publ., Kalamaroo—Michigan
Rowley G. (1959): Flowering Succulents, Farnham
Rowley G. (1978): The Illustrated Encyclopaedia of Succulents, Salamander Books, London

PERIODICALS AND SOCIETIES:

National Cactus and Succulent Jrnl., (NCSS), contact Miss W. E. Dunn, 43 Dewar Drive, Sheffield S7 2GR, England
Cactus and Succulent Jrnl. of GB, (CSSGB), contact CSSGB, 67 Gloucester Court, Kew Rd., Richmond, Surrey TW9 3EA, England
Cactus and Succulent Jrnl. of America (CSSA), contact Abbey Garden Press, Las Canoas Rd., Santa Barbara, Ca. 93105, USA

Nurseries Specializing in Cacti and Succulents:

England — Abbey Brook, Old Hackney Lane, Matlock, Derbyshire
—Whitestone Ltd., The Cactus Houses, Sutton-upon-Whitestonecliffe, Thirsk, N. Yorkshire Y07 2PZ
—The Exotic Collection, 16 Franklin Rd., Worthing, Sussex BN13 2PQ

USA — Abbey Garden, 4620 Carpinteria Ave., Carpinteria, Ca. 93013
— Epi Center, Box 1431, Vista, Ca. 92083
— Mesa Garden, 905 Impala Dr., Belen, N. M. 87002

INDEX OF SCIENTIFIC NAMES

Bold figures refer to number of illustrations.